THE
TOTALLY WIRED
WEB TOOLKIT

The
Totally Wired
Web Toolkit

Nathan J. Muller

McGraw-Hill
New York • San Francisco • Washington, D.C. • Auckland • Bogotá
Caracas • Lisbon • London • Madrid • Mexico City • Milan
Montreal • New Delhi • San Juan • Singapore
Sydney • Tokyo • Toronto

Library of Congress Cataloging-in-Publication Data

Muller, Nathan J.
    The totally wired Web toolkit / Nathan J. Muller : illustrations
by Linda Lee Tyke.
        p.   cm.
    Includes index.
    ISBN 0-07-044434-X (pbk.)
    1. Internet (Computer network).   2. Telephone systems.
3. Communications software.   4. Internet videoconferencing.
I. Title.
TK5105.875.I57M85   1997
004.67'8—dc21   96-38027
                                                                CIP

McGraw-Hill

*A Division of The **McGraw·Hill** Companies*

1 2 3 4 5 6 7 8 9 0   DOC/DOC   9 0 1 0 9 8 7

ISBN 0-07-044434-X

The sponsoring editor for this book was Brad Schepp, and the production supervisor was Pamela A. Pelton. It was set in Vendome by North Market Street Graphics.

Printed and bound by R. R. Donnelley & Sons Company.

McGraw-Hill books are available at special quantity discounts to use as premiums and sales promotions, or for use in corporate training programs. For more information, please write to the Director of Special Sales, McGraw-Hill, 11 West 19th Street, New York, NY 10011. Or contact your local bookstore.

 This book is printed on recycled, acid-free paper containing a minimum of 50% recycled de-inked fiber.

To my dear niece Amanda Lucio,
her husband Bernardo,
and their precocious little one, Jordan Ismaeil

CONTENTS

Contents

Contents

PREFACE

This book describes the various communications services that can be run over the Internet at little or no additional charge for your basic Internet access connection. The expense of an Internet connection can be recovered many times over by using the Internet for long-distance telephone calls, faxing, e-mail, paging, voice messaging, conferencing, and collaborative computing. Since the Internet is international in scope, the savings can accumulate fairly rapidly. All that is needed is the right hardware and software, which can be purchased at very modest cost.

In the case of software, much of it can be downloaded free over the Internet, either as shareware, freeware, betaware, or limited-use trialware. And if you've been looking for an excuse to buy a sound card for your computer, look no further! Aside from your modem, the sound card is the key hardware component that will enable you to fully exploit the power of the Internet. Like modems, sound boards are now bundled with most new computers, but they can also be purchased separately through mail-order firms for as little as $125. But the cost of hardware and software is paltry compared to the benefits you will enjoy when they are used for communicating with others worldwide over the Internet.

This book is intended for a broad audience of experienced Internet users, as well as relatively new Internet users. Even those who are not yet connected to the Internet will find this book useful. In learning about the range of money-saving communications opportunities the Internet offers, they will have all the incentive they need to get online as quickly as possible to take advantage of them. For those who are tired of just browsing the Web, this book will be of special value.

Browsing the Web has become a familiar pastime for many people who have subscribed to an Internet access service in recent years. The term is apt as it is worn: It suggests aimless wandering, nomadic restlessness, and endless searching for something that we hope will engage us, even if only briefly.

For all the hype about the Web, there is really very little that merits our attention and even less that is relevant to our interests. In fact, the number of sites that offer useful information of any kind is too few and far between. Even discussion groups that supposedly cater to specialized interests seem dominated by cloddish people who get an ego boost by flaming and otherwise belittling others who are not as technically astute or knowledgeable about a topic as they. The apparent anonymity and distance that the Internet affords allows people to say things they wouldn't dare utter in face-to-face conversation.

The situation seems to be getting worse, rather than better. As more people discover the Internet, the more it becomes cluttered with useless information. Too many Web pages proudly proclaim that they are still "under construction," as if the rest of us are waiting in breathless anticipation. Many have nothing better to offer than a long list of links to other Web sites, which supposedly are the "really cool" places to be. Other Web pages are mere garbage dumps for pornographers, hate groups, con artists, and the merely self-absorbed.

There is worthwhile information on the Internet, to be sure, but it must be ferreted out with dogged persistence. Web browsing is a novelty that quickly runs its course, even if you load the latest and greatest browser software every six weeks. Eventually one is left wondering what all the fuss is about.

The World Wide Web originated as a way to publish and share research papers among scientists and academicians, making documents easy to find and access. For most of us, however, this model is not applicable, yet we seem wedded to it as if it were. This may help explain why there is so much junk on the Web—everyone seems compelled to publish something, no matter how trivial, even if it is only a list of other Web sites.

For most of us, the real value of the Web is not in being able to publish and distribute important information worldwide, but in empowering ourselves through the use of various forms of audio and visual communications that are accessible through the Web. With the right tools, we can engage in activities that are immensely more worthwhile and beneficial in our personal and business lives than merely browsing the Web.

If you are self-employed and working at home, the Web can become a valuable business tool, allowing you to send e-mail, faxes, and messages to anyone, anywhere—even if the recipient does not have an account with an Internet service provider. Over the Internet, you can make phone calls to business associates and partners, saving hundreds of dollars a month in long-distance charges. You can even set up a videoconference, view documents and take turns marking them up, and share applications over the Internet.

If you *telecommute*—work at home most of the time instead of at a corporate office—the Internet can be the lifeline that keeps you in touch with colleagues and connects you with various corporate databases. In fact, virtually everything you do at the corporate office can now be done from home via the Internet—even meeting others at a virtual water-cooler! And, since being able to do all these things requires only an Internet access account costing between $10 and $20 a month, companies have more incentive to try flexible work arrangements such as telecommuting.

Think about the ramifications of telecommuting to your personal situation. It allows you to balance work and family life, which in itself has so many advantages. The arrangement allows you to supervise children instead of shipping them off to day care (sick or not) or leaving them home alone after school to fend for themselves. It gives you the flexibility during the day to meet with teachers, keep appointments with the doctor, renew your driver's license, and attend your kid's baseball game. It allows you to work at your peak energy level instead of arriving at the office exhausted from a long commute. Working at home also allows your work ethic to rub off on young children, better preparing them for college and career. Finally, the telecommuting arrangement can reduce stress, allowing you to lead a healthier, happier life. All this is possible simply by taking advantage of the Internet.

This book will reveal these possibilities to you through a discussion of the various technologies and applications—many of which have only recently become available. You will learn how to use these tools on the Internet for enhancing your ability to communicate in a variety of ways. In the case of telephone calling, faxing, or videoconferencing over the Internet, substantial cost savings on long-distance charges can be achieved. If you already have a home page on the Web, this book shows you how to voice- and video-enable it to achieve a new level of interactivity. The book also shows you how you can meet with business colleagues and partners on the Internet and work together collaboratively to complete projects and meet deadlines.

This book will also be of value to the vast majority of non-Internet users as well, since it provides compelling incentives to finally get on the Internet. Many of these people see no need to spend a lot of money on a computer and Internet access connection just to "surf the Web," but will appreciate being able to communicate in sophisticated ways with friends, relatives, and colleagues anywhere in the world—things they may not have considered doing before because of high telephone charges, especially if they are students, military, or retired people on low or fixed incomes.

The homebound elderly will especially appreciate the vast new world of possibilities that are suddenly opened to them via the Internet. They can visit with relatives and friends through a videoconference over the Internet or place telephone calls without considering the expense. They can take part in chat rooms devoted to topics of interest to them. Instead of feeling left out of family and community life, due to illness or frail condition, they can use the Internet as a means to participate with others and, in many cases, regain a sense of identity and purpose.

If you're among the millions of people with a computer and Internet connection, it's up to you to absorb what's in this book and get these tools into the hands of relatives and friends who are less fortunate. You will not only contribute to their quality of life—which in itself is a good thing to do—but in doing so you will expand your own opportunities for learning, teaching, and helping. These activities have their own rewards.

It is not the intent of this book to provide an exhaustive description of all the products and services that can be used on the Internet to save on communications costs, but only to provide examples that illustrate various concepts. As such, this book provides a good foundation for further research into specific products and services. Whenever possible, the Web pages of vendors and service providers are noted. You are encouraged to periodically check these pages for updated product and service information.

For your convenience, you can go to the following Web page for a listing of all vendors mentioned in this book, including links to their home pages where you can download free software for use or evaluation:

http://www.ddx.com/mgh.shtml

Here, you will even find updates and new material on topics related to the content of this book.

The information contained in this book, especially as it relates to specific vendors and products, is believed to be accurate at the time it was written and is, of course, subject to change with continued advancements in technology and shifts in market forces. Mention of specific products and services does not constitute an endorsement of any kind by either the author or the publisher.

—NATHAN J. MULLER

THE
TOTALLY WIRED
WEB TOOLKIT

Chatting Over the Internet

Introduction

One of the most attractive features of the Internet is the ability to use it for real-time chat. *Chat* allows users to converse with each other interactively in text mode via the computer keyboard. This interactivity distinguishes chat from other text-based services such as newsgroups and electronic mail, which only allow messages to be posted for access at a later time. Anything you type during a chat session will be seen by all the other participants on that channel, who can respond immediately with comments of their own.

You can use chat to make new friends, engage in lively debates on current issues, or communicate privately with friends and relatives anywhere in the world. You can chat with movie stars, music groups, sports figures, and other celebrities who host chat sessions. Companies also host chat sessions, giving people from around the world an opportunity to ask questions about the products or services they offer.

Companies such as Nintendo and Sega hold chat sessions in which they take input from customers to help them develop new games. Many writers and authors hold chat sessions to talk to their readers and promote new material. In addition, many companies offer a chat capability in conjunction with their private Web sites, allowing employees to discuss customer problems and solutions.

You can single out one user to chat with or chat with several users at once, similar to the way you can communicate person-to-person on a telephone or with a number of people through a conference call. The only differences are that you are using the Internet and you are typing your end of the conversation instead of talking. If you do not want to use your real name, you can use a nickname instead. You can even make yourself invisible to other users as you monitor a selected channel.

Chat Environments

There are several environments in which chat is used.

- *Internet Relay Chat (IRC)*.   This is a pure chat system and the main IRC network often has tens of thousands of users online at any given time. You can access the IRC network and participate in scheduled or ongoing chat sessions that are organized around themes or topics.

- *World Wide Web.*   Generally, the Web is a static medium permitting the one-way transfer of data. However, with the integration of chat on the Web, people who have a common interest can meet at a Web page from which they can launch a chat application to converse with others. The addition of Web-integrated chat enables the Web site to develop a sense of community among its visitors who already have a common interest in the content of the site.

- *Telnet.*   Some Internet sites are only accessible via the Telnet protocol, which provides basic connectivity between two computers. This connection enables you to type commands which are executed at the host. The results are passed back to your computer. Among other things, these hosts often include real-time chat systems and games. If you do not have your own IRC chat software, you can access the IRC network from a Telnet host. However, there will be more delay and user restrictions than if you had your own software.

- *Point to point.*   There are chat programs that allow you to communicate with another person directly, without the need to meet at an IRC or vendor-supplied server. All you need to establish a point-to-point connection is the other person's e-mail address. Many chat programs offer this feature, in addition to supporting the IRC protocol.

- *Multi-User Dungeons (MUDs).*   These are chat systems that are used mostly for games. They typically include graphical components that can be manipulated within a three-dimensional fantasy world. There are over 500 MUDs, many with thousands of registered players worldwide. A hundred or more players may be active in the same MUD at any given time.

Some chat programs are based on the MUD concept. Users are represented with animated graphic images called *avatars*. Participants meet in three-dimensional worlds by moving their avatars like game pieces, which users control with the mouse or keyboard. Users can select an avatar from a gallery provided with the software or draw their own avatar. One such program, Worlds Chat, provides a space station that can be explored by the avatars. As you encounter other avatars, you can interact with them by using the keyboard to chat. A Whisper feature allows you to chat privately with one person, in which case, other participants will not be able to see what you type.

In the past, participating in chat sessions involved logging into a UNIX account and using arcane keyboard commands. The graphical user interfaces (GUIs) of today's Windows and Macintosh products greatly simplify access to chat networks and their use. The most frequently used com-

Figure 1.1

The pull-down command menu in the mIRC 4.1 chat program.

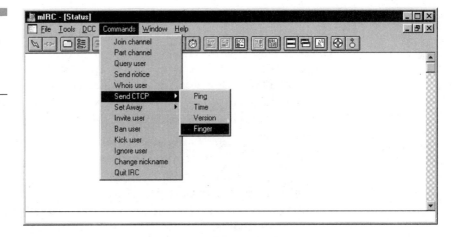

mands can be picked from a pull-down menu (Fig. 1.1). There may still be some arcane commands to learn, but even these can be simplified by setting up aliases (discussed later) which allow you to replace these commands with ones that are easier for you to remember.

The GUIs of some chat programs, such as Quarterdeck's Global Chat, include a banner feature that displays a series of graphic presentations, usually advertisements, at the top of the chat window (Fig. 1.2). This dis-

Figure 1.2

Quarterdeck's Global Chat interface.

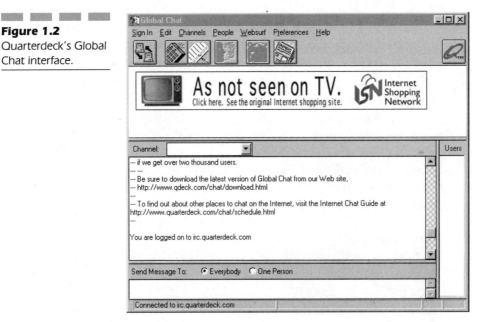

play works only when you connect to a chat server and the content is controlled by the particular server you are connected to. The banners themselves usually contain hypertext links, so that when you click on the banner, a Web browser will jump to the foreground and automatically connect you to the URL associated with the company or service in the banner. When you connect to a chat server that does not support the banner feature, a static banner fills the display space.

More chat products are starting to support audio and speech, allowing you to use commands that cause audio files to be played at the remote computer or even allow participants to engage in voice conversations. One product, InterFACE from HiJiNX in Australia, even allows you to edit your voice with a special voice editor. This allows you to modify the voice you want people to hear when they are chatting with you. InterFACE includes a selection of preset voices in case you would rather not tinker with the technical details.

Software and Hardware Requirements

All you need to participate in online chatting is an account with an Internet service provider (ISP), a TCP/IP dialer, and the chat software. Most chat software is shareware or freeware and can be downloaded to your computer from various bulletin board systems (BBSs) and Web sites.

The most basic piece of software you will need is a TCP/IP* dialer program, which enables packet mode operation so you can communicate over the Internet via the modem. If you already have an Internet access account, you're probably using this software to log on to the Internet. Nevertheless, TCP/IP dialers are available as low-cost shareware. One of the most popular and easy to use third-party TCP/IP dialers is Trumpet, which can be downloaded for a 30-day evaluation from numerous Internet and BBS sites (Fig. 1.3).

A very good TCP/IP dialer comes with Windows 95 (Fig. 1.4) and you can set up multiple instances of the dialer, each with a different phone number.

---

* Without getting technical, TCP/IP stands for Transmission Control Protocol/Internet Protocol. This is a suite of protocols and services that handle data over the Internet. Further discussion of TCP/IP is beyond the scope of this book and, for the most part, is unimportant to most Internet users.

Figure 1.3
Trumpet Winsock
dialer interface.

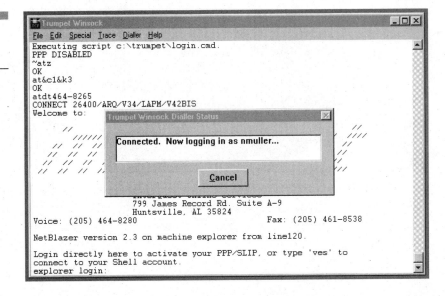

Chat programs will run on most types of Windows and Macintosh computers. The minimum memory requirement is usually 4 MB, but realistically you will need 8 MB of RAM to ensure adequate performance. However, with today's graphically enriched applications, you should have 16 MB of RAM to do everything explained in this book, plus run your existing applications.

Obviously, your computer will need a modem to establish and maintain the Internet connection. A modem operating at 14.4 Kbps is the minimum requirement, although a 28.8 Kbps modem would be better. If you are thinking of buying a modem, you can purchase a 28.8 Kbps modem card from a mail-order catalog for around $120.

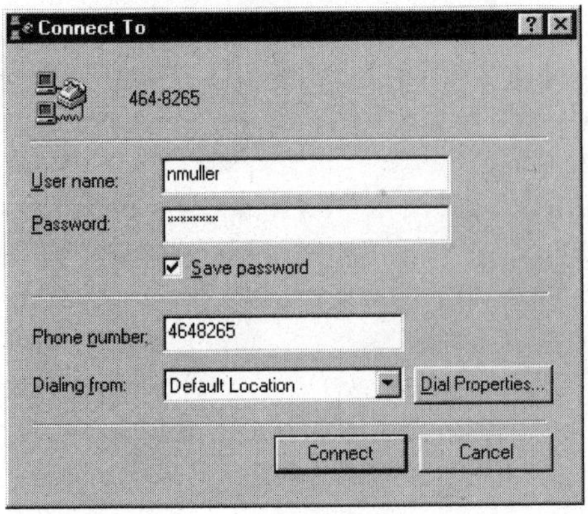

Figure 1.4
Windows 95 TCP/IP
dialer interface.

Although a sound card is not required to chat over the Internet, keep in mind that these programs are continually being enhanced with new audio and real-time voice features. You will need a sound card to take advantage of these features plus use the other applications described in this book, such as making telephone calls over the Internet (Chap. 2) and videoconferencing over the Internet (Chap. 3). A good sound card can be purchased through mail order for around $150. A hardware bundle that includes a sound card, plus a pair of speakers and a self-powered sub-woofer, can be had for about $225.

Internet Relay Chat

Internet Relay Chat (IRC) has been around since 1988 and is the global standard for chatting over the Internet. You need IRC client software installed on your computer to connect to an IRC server. The servers pass messages from user to user over the IRC network. Once connected, you will be able to chat with users who are accessing not only the server you are connected to but also users on any of the servers that are on the same network.

Discussions over IRC take place in channels. A *channel* is created when the first person names it. While the channel exists, anyone can reference it by name to join or leave it. Channel names can be up to 200 characters in length and must begin with either the & (ampersand) or # (pound) symbol. The only other restriction on a channel name is that it may not contain any spaces. This is usually overcome by using an underscore (_) between words in a channel name.

Depending on the server you connect to, there may be a few dozen or a few thousand channels open at any given time. Each channel usually has a theme or topic that draws together individuals with the same interest. However, nothing forces users to follow the suggested topics except that their comments may go ignored. If off-topic comments become annoying, the user can be kicked off the channel by the channel operator.

You may have run across so-called *chat rooms*. These are simply servers or channels where users meet to chat with each other. In other words, a chat room is just a common place on the Internet where users can go to communicate with people who share their interests. More sophisticated chat rooms are graphical in nature, allowing users to move avatars and other objects through an online shopping mall or fantasy world.

The commands used on the IRC network start with a slash (/), and most are one word. For example, typing */HELP* will get you help information and typing */NAMES* will get you a list of channels currently active on the server. Typing */LIST* will also get you a list of channels, but will also include more details, such as the number of people on the channel and the topic of discussion. There may be commands that are specific to the chat software you are using.

To get started with a chat session, you dial the Internet access number provided by your Internet service provider (ISP) using your favorite TCP/IP dialer program. After logging on to the Internet with your user name and password, you simply open the chat program. When using chat software for the first time, you will need to access a setup screen (Fig. 1.5) that allows you to enter some basic information such as your real name, e-mail address, a nickname, and an alternate nickname. You can also choose the server you want to connect to from among a list of IRC servers in different countries. The IRC networks of DALnet, EFFNET, and Undernet offer the most servers. Establishing a connection is as simple as clicking on a server from the list.

If your IRC client software does not provide a list of servers to connect to, you can do a search on the Web for *IRC servers* and compile your own list of servers that are nearest to your location. If you want to get started right away, you can try the following Undernet servers in North America:

albany.ny.us.undernet.org	128.213.5.17	New York
austin.tx.us.undernet.org	128.83.162.106	Texas
boston.ma.us.undernet.org	129.10.22.11	Massachusetts

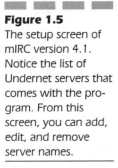

Figure 1.5
The setup screen of mIRC version 4.1. Notice the list of Undernet servers that comes with the program. From this screen, you can add, edit, and remove server names.

davis.ca.us.undernet.org	128.120.2.8	California
manhattan.ks.us.undernet.org	129.130.8.12	Kansas
milwaukee.wi.us.undernet.org	140.104.4.169	Wisconsin
norman.ok.us.undernet.org	129.15.22.33	Oklahoma
pasadena.ca.us.undernet.org	131.215.99.9	California
rochester.mi.us.undernet.org	141.210.10.117	Michigan
sanjose.ca.us.undernet.org	192.160.13.4	California
stgeorge.ut.us.undernet.org	144.38.16.2	Utah
tampa.fl.us.undernet.org	131.247.31.19	Florida
washington.dc.us.undernet.org	152.163.51.22	Virginia

When the connection is made to the IRC server (or some other chat server), typically you will be greeted with the *MOTD* (message of the day) from the server administrator informing you about any problems on the network and what alternative servers might be available. There is usually other useful information such as the number of people currently connected, how many channels have been formed, and how many people are invisible (Fig. 1.6).

Once you are logged on to the IRC network and join a channel, anything you type will be seen by all the users on that channel. If a particular channel does not exist, one will be created for you. By default, you are

Figure 1.6

The message of the day (MOTD) from MIT's IRC server as viewed in Winsock IRC 3.1.

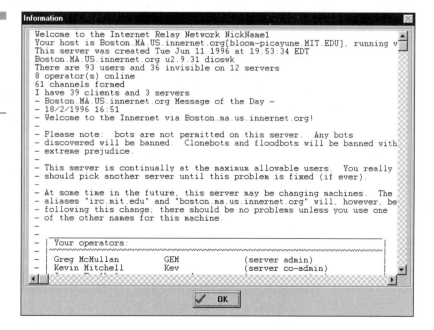

```
Information                                                              ☒
  Welcome to the Internet Relay Network NickName1
  Your host is Boston.MA.innernet.org[bloom-picayune.MIT.EDU], running v
  This server was created Tue Jun 11 1996 at 19:53:34 EDT
  Boston.MA.US.innernet.org u2.9.31 dioswk
  There are 93 users and 36 invisible on 12 servers
  8 operator(s) online
  61 channels formed
  I have 39 clients and 3 servers
  - Boston.MA.US.innernet.org Message of the Day -
  - 18/2/1996 16:51
  - Welcome to the Innernet via Boston.ma.us.innernet.org!
  -
  - Please note:  bots are not permitted on this server.  Any bots
  - discovered will be banned.  Clonebots and floodbots will be banned with
  - extreme prejudice.
  -
  - This server is continually at the maximum allowable users.  You really
  - should pick another server until this problem is fixed (if ever).
  -
  - At some time in the future, this server may be changing machines.  The
  - aliases "irc.mit.edu" and "boston.ma.us.innernet.org" will, however, be
  - following this change; there should be no problems unless you use one
  - of the other names for this machine.
  -
  - | Your operators:
  - |
  - | Greg McMullan          GEM          (server admin)
  - | Kevin Mitchell         Kev          (server co-admin)

                          [ ✓  OK ]
```

the channel operator. If your channel is of interest to others, they will join you there for a discussion.

Commonly Used Chat Commands

Once you join a channel you can participate in the ongoing discussion simply by typing text on the command line (Fig. 1.7) and then hitting the return key. You can ask a question, answer another person's question, or offer your opinion or insights on any matter being discussed. Just remember, anything you type from this point on will be seen by all participants on that channel.

Once you are connected to a channel or chat room, there are a number of simple commands you can use to facilitate participation. The command list below is not intended to be exhaustive, but it will enable you to get started right now. As you gain more experience with this form of communication, you will pick up shortcuts and advanced techniques from the people you chat with.

The following IRC chat commands are not case sensitive; you can type them in either upper- or lowercase.

Figure 1.7
The command line in MeGALiTH's Visual IRC (beta version). Notice the command line at the bottom of the screen. This is where the user types in commands. In this case, the user wants to join a channel named teapot. All chat programs have a command line like this.

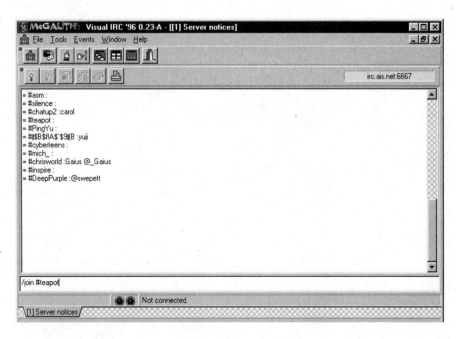

/LIST

Once logged on to a particular server, you will probably want to start by finding out what channels are currently operating so you can decide what discussion to join. The /LIST command will retrieve a list of all channels active on the IRC server you have logged on to. Because there may be thousands of active channels, you may want to limit the list by the number of participants or by topic of interest.

For example, the command *LIST -MIN 10* will provide a list of only the channels having a minimum of 10 participants (Fig. 1.8). The command *LIST -MAX 10* will provide a list of only the channels having a maximum of 10 participants. You can combine these parameters within a single command, such as *LIST -MIN 10 -MAX 20,* which will provide a list of only the channels having at least 10 participants but no more than 20 participants.

You can also obtain a list of channels that have a key word in their name. The command *LIST #Windows95,* for example, will retrieve a list of channels in which Windows 95 is the main topic of discussion.

/JOIN

Once you have obtained a list of available channels, you can select one to join. This is done with the command /JOIN. If you have selected a chan-

Figure 1.8
On the command line of mIRC 4.1, a list of channels having at least 30 participants is requested with the command /LIST -MIN 30 entered on the command line (see background screen). According to the list that is returned, 42 of the 3855 active channels fit that criteria.

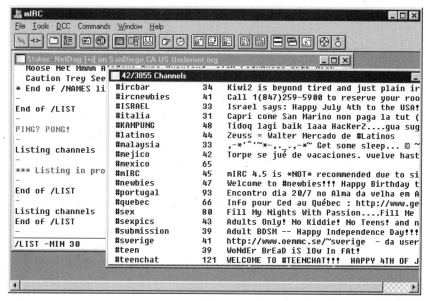

nel named Windows95, for example, the command will be */JOIN #Windows95.*

When you join a channel, you will probably get a welcome message from the channel operator, which is the same message used to greet everyone else who joins the channel. Your name or nickname may even be echoed back within the context of the message, making the greeting seem more personal.

You can easily create your own channel. To do this, just type */JOIN #channelname,* where channelname is the name of the channel you want to create. If there is no channel already using that name, your channel will be set up automatically and you are in charge of it. If you join a channel and you see that only your name is listed, you just created that channel. This means that you can set the operating mode of the channel. (See the section on the /MODE command that follows.) The channel disappears when the last person leaves it.

At least one IRC network, DALnet, allows users to register their channels. When a channel is registered, nobody else can create a channel with the same name, even if it is not in use at the moment.

/DESCRIBE

If you have your own channel, you can send a description of it to a specified person without the rest of the channel seeing the message. For example, the command can be used in the following sense:

/DESCRIBE NetDog Newbie This channel discusses tips and tricks to help you get the most out of Windows 95.

In this case, you would be sending a description of your channel to a person with the nickname NetDog who is currently on the channel named Newbie.

/INVITE

If you spot someone you would like to chat with, you can send an invitation to that person. For example, the command */INVITE NetDog Windows95* is an invitation to the person nicknamed NetDog to join you on the channel named Windows95. If no channel name is provided in the command, it is assumed to be the current channel you issued the invitation from.

/KICK

If you are the channel operator, you can kick a specified user off your channel using the /KICK command. This can be used to kick out abusive users, as in */KICK NetDog*. This command can also be used to kick off users who flood your channel with useless information.

/TOPIC

If you are the channel operator, you can change the topic of conversation to anything you wish by using the /TOPIC command. For example, let's say your nickname is NetDog and you want to change the topic of your channel from Windows95 to WindowsNT. The command would be written as follows:

/TOPIC #Windows95 WindowsNT

Users on your channel will see the change as:

*** NetDog has changed topic to "Windows NT"

/MSG

You can engage in private conversations with other people without leaving the channel or opening a separate query window. For example, the command */MSG NetDog* will notify the person with the nickname Net-Dog that you wish to chat privately. If that person agrees, the discussion will take place without other channel participants being able to see what is being typed between the two of you.

If you want to send a private message to more than one person, you can specify a list of nicknames separated by commas (no spaces), followed by the message. For example, */MSG NetDog,NetKitten,NetMouse Hey, what are you folks doing tonight?*

/QUERY

This command is similar to the /MSG command previously discussed and is used in a similar manner except that it opens a query window to the user when a private message is sent, allowing longer two-way conversations without having to type /MSG before everything you say. Here's an example:

> */QUERY NetDog Hi! You can check out my Web page at www.ddx.com for a*
> *white paper on that topic.*

This opens a query window to the user whose nickname is NetDog. The window will contain the message about the white paper and anything else you say after that. When you want to end the conversation and rejoin the group, you just type /QUERY without anything else after it.

/ME

With the /ME command you can send *action* messages that convey some type of physical reaction when words fail you. For example, NetDog might respond to something that is said with the following command:

> /ME has had enough, let's move on to something else!

Other participants in the discussion will see the message as:

> *** ACTION NetDog has had enough, let's move on to something else!

/SOUND

Some chat software—PIRCH, for example—allows you to cause a multi-media file to be played on another person's computer system. The /SOUND command can be sent to an entire channel or to a select individual on a channel. The components of the command are as follows:

> */SOUND #channelname or nickname filename message*

where *filename* is the name of the multimedia file and *message* is the text that you want to accompany it.

PIRCH supports several media file types for use with the /SOUND command, including:

.wav   Standard Windows Wave file format for sounds

.mid   Midi file format

.avi   Video file format

When a video file type is played, it will be displayed in a separate pop-up window. If no file extension is included on the filename, PIRCH will assume that it is a .wav file and append the extension automatically. The multimedia file itself is not transmitted through the IRC network; in order for the command to work, the file must exist on the other person's computer system.

/WHOIS

If you would like to know more about a particular person you encounter on the same channel, you can request additional information by using the /WHOIS command. For example, */WHOIS netdog*, will provide additional information about the person who goes by the nickname NetDog.

The information you get back usually includes the person's real name and e-mail address, which is entered in the software's setup window. Since you see only the information that person wants you to see at any particular time, you should be aware that it could be false information. This command will also provide you with the server through which the person is connected.

/WHOWAS

With this command, you can ask for information about a nickname which no longer exists due to a nickname change or the user leaving the IRC network. In response to this query, the server searches through its nickname history, looking for any nicknames that are an exact match (no wildcards are allowed for nicknames). The history is searched backward, returning the most recent entry first. If there are multiple entries, the user-specified number of entries will be returned. If no number of entries is specified, all of them will be returned. The following examples show how the /WHOWAS command can be used:

/WHOWAS NetDog returns all information in the nickname history about all instances of NetDog.

/WHOWAS NetDog 7 returns, at most, the seven most recent instances of NetDog.

/WHOWAS NetDog 1 *.com returns the most recent history for NetDog from the first server found to match .com.

/AWAY

If you would like to take a break and remain on the channel, you can leave an appropriate message explaining that you are not currently available. When someone invites you to talk privately, that person will get your message. The /AWAY command, followed by a short message, is used for this purpose, as in the following example: *AWAY I'm looking for a snack. I'll be right back!*

When you return, you simply type the command /AWAY with no message. This will turn off the message and enable you to accept queries from other participants on the channel.

/NOTIFY

Sometimes you will want to be notified of when certain people enter or quit the channel. For example, you can arrange a time and place to meet for a chat with your son or daughter who is away at college or in the military at an international location. Phone calls between international locations can be prohibitively expensive. However, through e-mail you arrange to meet at a specific time at a specific IRC server. Then you use the /NOTIFY command to be alerted of when that person enters the server so you can arrange a private discussion using the /QUERY command. For example, *NOTIFY NetDog* will let you know when the person with the nickname NetDog enters the IRC server. When NetDog logs on to the server, you will be alerted with a message that looks like this:

*** Signon by NetDog detected

If NetDog is already logged on to the server when you issue the /NOTIFY command, you will receive a message that looks like this:

*** Currently present: NetDog

/IGNORE

There may be occasions when you do not want to receive messages from particular users. Perhaps they are harassing you, sending you objectionable material, or just talking trash. You can't prevent these users from sending out such messages, but you can prevent them from being displayed on your computer screen by using the /IGNORE command. You can suppress output by nickname or e-mail address or by message type. You can even use wildcards to suppress a wider range of people you do not wish to have contact with.

To suppress a message from a particular nickname, the command is used as follows: *IGNORE NetDog.* Whenever a person with the nickname

NetDog tries to contact you, the message will be suppressed. To suppress a message from a particular e-mail address, the command is used as follows: */IGNORE nmuller@ddx.com.* To suppress messages by type of message (i.e., MSG, QUERY, INVITE, etc.), the command is used as follows: */IGNORE QUERY.* In this case, whenever anybody tries to initiate a private conversation with you using the /QUERY command, those messages will be suppressed. The sender will receive a message indicating your refusal to accept.

Wildcards may be used in all formats. For example, */IGNORE *@ddx.com ALL* will suppress all messages from users having an e-mail address that ends with ddx.com, in which case the sender gets back a message that looks like this:

*** Ignoring ALL messages from *@DDX.COM

To remove nicknames, e-mail addresses, and message types from suppression so that you can once again receive messages, the /IGNORE command is used in the following way: */IGNORE *@ddx.com NONE.* In this case, you will once again be able to receive private messages from anyone with ddx.com in his or her e-mail address.

/NICK

If the nickname you selected is already in use by someone else, you will not be able to enter the IRC server until you change it. You can change your nickname at any time with the /NICK command as follows: */NICK CommDog.* Then you will see a message that confirms the change, such as:

*** NetDog is now known as CommDog

On IRC, users are identified by their nickname. Most IRC nets limit these to nine characters in length, but at least one IRC net, DALnet, allows nicknames of up to thirty characters long to allow for greater depth and originality when choosing a nickname.

Generally, there is no ownership privilege on the use of nicknames. You can request that people stop using your nickname, but they do not have to comply with your wishes. There is no enforcement authority to help you. Even though IRC server operators and channel operators have the power to kick people off their facilities, they generally will not get involved in sorting out the use of nicknames.

Again, DALnet is the exception to this rule. It allows users to register their nicknames and will not allow other users to commandeer someone else's registered nickname.

/PART

To leave a channel, the /PART command is used, along with the name of the channel. For example, to leave the channel Windows95, you would type */PART #Windows95*. This command lets you leave channels without disconnecting from the IRC network so you can join other channels. You could also use the /LEAVE command for this purpose.

/QUIT

To disconnect from the IRC network, you use the /QUIT command. You have the option of stating the reason for leaving, as in: */QUIT It's getting late, bye!* This message will only be seen by participants on the same channel as you.

/MODE

If you have set up your own channel with the /JOIN command, you are the channel operator and can set the operating mode of the channel. This is done with the /MODE command as follows:

/MODE [channelname] +1.   This allows you to specify the maximum number of users that are allowed on your channel. Just change the number 1 to the desired number of users.

/MODE [channelname] +b [nickname] or [user address].   This allows you to ban someone from your channel by specifying that person's nickname or user address.

/MODE [channelname] -b [nickname] or [user address].   This allows you to remove the ban on someone by specifying that person's nickname or user address.

/MODE [channelname] +i.   This allows you to make yourself invisible to anybody that does not know your nickname. In other words, only people who know your nickname will be able to find you using a search method. However, not even searches that use a wildcard to approximate your nickname will reveal your whereabouts on the IRC server.

/MODE [channelname] +m.   This allows you to set up your channel as *moderated,* meaning that only you and your designated channel operators can talk. Everyone else can only monitor the conversation.

/MODE [channelname] +n. This allows you to prevent external users from engaging in private conversations with people on your channel.

/MODE [channelname] +o [nickname]. This allows you to give someone else operator status, in case you want to leave or take a break. This confers on that person the ability to change the operating mode of your channel.

/MODE [channelname] +ooo [nickname1][nickname2][nickname3]. This allows you to give operator status to several people.

/MODE [channelname] -o [nickname]. This allows you to take back operator status from someone.

/MODE [channelname] +p. This allows you to designate your channel as *private,* meaning that nobody else can join. This is useful for meeting friends, family, or colleagues for private conversations.

/MODE [channelname] +s. This allows you to designate your channel as *secret,* meaning that it will not show up in a list of channels. Participants' names do not show up in a names list of people on IRC. This provides an added level of privacy when friends, family, or colleagues meet on a channel for conversations.

/CLEAR

Finally, a handy command to have at your disposal is the /CLEAR command. This command clears your screen, allowing you to eliminate all the clutter you have accumulated through various chat sessions so you can refocus your efforts.

Creating Aliases

Aliases are custom commands that you can define for your own use, but which are implemented on an IRC server in their standard form. They are simply short, easy-to-remember substitutes for longer commands. In order to make use of aliases you must of course already be familiar with the basic IRC commands. If you happen to create an alias with the same name as an existing IRC command, the chat program will give precedence to IRC commands over aliases.

Not all chat programs support aliases. Those that do, provide an alias definition window where the alias and its definition are entered on a single line. The simplest aliases include only the alias name followed by the command to be executed. In PIRCH, for example, *P:/PART #* allows you to leave the current channel simply by typing the alias. */P.WIN95:/JOIN #Windows95* allows you to join the Windows95 channel simply by typing the alias /WIN95. PIRCH also allows you to define the function keys F1 to F12 on your keyboard as aliases. To prevent conflicts with other programs such as TSRs (terminal stay residents), the function keys can be used in combination with either the Alt or Ctrl key or used with both. For example, you can assign F2 for the command */AWAY I need to take a break, but continue without me—I'll catch up when I get back.* However, if F2 is already in use by another application to call up a notepad, for example, you will have to assign Alt-F2 or Ctrl-F2 for the command alias.

There are some function keys that are reserved by Windows, and the IRC chat software will make no attempt to override their default definition. For example, Ctrl-F4 closes the active window, while Alt-F4 closes the application. Although not a Windows-reserved function, F1 calls up help within most Windows applications. Therefore, when creating aliases these key combinations should be avoided.

Non-IRC Chat

As noted earlier, there are alternative ways to chat over the Internet that do not necessarily involve IRC servers. With PowWow, for example, all you need to establish a connection with someone is that person's e-mail address. PowWow has many of the same features found in IRC chat programs, but calls them by user-friendly names. PowWow's answering machine, for example, corresponds with IRC's /AWAY command, which allows you to leave a message when people try to contact you while you're away from your computer. PowWow's call-blocking feature corresponds to IRC's /IGNORE command, which allows you to refuse chat requests from other people.

PowWow offers two conferencing modes. Personal Communicator Mode allows up to seven people to chat together as a group. Participants can chat via keyboard, voice, cruise the Web, play audio files, and transfer files. In this mode, text is *streaming*—sent immediately to the other people they are chatting with keystroke by keystroke.

PowWow's Conference Mode allows up to 50 people to chat together as a group or listen to a speaker or speakers and view the moderator's Web page links. Text is sent in *block mode,* that is, only when the Enter key or

Send button is pressed. If the conference is moderated, the person running it can display pages on each attendee's Web browser as well as selectively allow people to talk. Conference Mode can be used for providing online presentations, lectures, training, and other discussions. A list of PowWow conferences that are available 7 days a week, 24 hours a day can be found at:

http://www.tribal.com/conferences/default.htm

You can add your conferences to the listing by filling out the form at:

http://www.tribal.com/conferences/confadd.htm

If you try to page someone who is connected to the maximum number of users, or if joining them would bring you over the maximum number of users, the connection will be refused.

You can find other people to chat with by checking the Tribal Voice PowWow White Pages server. There, you can find other users to chat with or add yourself so that others may find you in the future. The White Pages is organized by user name, geographic location, interests, or by listing the last 10, 25, or 50 people who have run PowWow. To search the Pow-Wow White Pages server, go to the following URL (Fig. 1.9):

http://www.tribal.com/wpsearch.htm

Figure 1.9
PowWow's White Pages allow users to find each other through a search that can include a person's name, Pow-Wow ID, city, state, country, or area of interest.

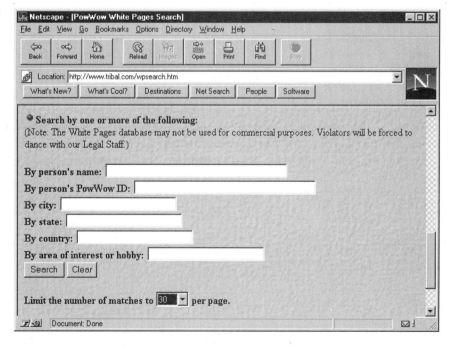

A fun feature offered by PowWow is the ability to cause sounds to be played when certain events occur or to accompany text sent to another person during a chat session. PowWow comes with the following sound files:

APPLAUSE.wav which plays applause and clapping

BOO.wav which plays "Boo!"

BYE.wav which plays "Good-bye" (male voice)

BYE_F.wav which plays "Good-bye" (female voice)

CHATREQ.wav which plays "You have a chat request"

COOL.wav which plays "Cool" (male voice)

COOL_F.wav which plays "Cool" (female voice)

HI.wav which plays "Hi there" (male voice)

HI_F.wav which plays "Hi there" (female voice)

LAUGH.wav which plays laughter

LOON.wav which plays a PowWow alert when events occur within PowWow

STILLTHR.wav which plays "Are you still there?"

VOC_BGN.wav which notifies you to begin voice chat

VOC_BTN.MID which accompanies use of the voice chat button

VOC_END.wav which notifies you of the end of voice chat

PowWow users have the option of turning off the ability to play sounds on their computer in response to commands issued by the people they chat with.

PowWow also offers a simple whiteboard (Fig. 1.10) that allows you to share drawings with the people you chat with. It includes many of the standard tools found in graphic design products, including a pencil for freeform drawing, a point-to-point line drawing tool, and a brush tool. There are also tools for drawing shapes such as rectangles and ellipses, as well as for adding text to drawings. Other functions include copy, undo/redo, and clear. You can even call up a list of users who are viewing the whiteboard.

Features

Chat products differ in terms of performance and features. After narrowing down your choices to a few products based on how well the vendors' evaluation software performed, you can turn your attention to compar-

Figure 1.10
PowWow's white-board offers a number of tools for drawing shapes, filing them in, and adding text. Controls are available for changing line widths, text font and point size, and adding color.

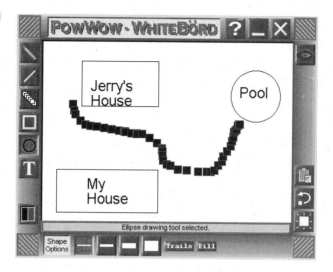

ing features to make a final selection. Here is a list of features you may want to look for, especially when choosing Windows-based products:

- *Address book.* Allows you to set up a directory of people you chat with frequently. The addresses can easily be added, viewed, and removed.

- *Alias support.* Provides the means to turn long commands into shorter, easier-to-remember commands. This function can also be used to generate standard replies, jokes, or actions.

- *Audio conference.* Allows you to speak as well as chat with other users.

- *Browser plug-in capability.* Some chat programs can be plugged into Web browser software such as Netscape Navigator, allowing you to connect to chat servers around the world by clicking on a link rather than having to launch a separate application. Netscape Communications even offers a chat program called Netscape Chat (Fig. 1.11).

- *Channel filtering.* You can search for one or more words matching channel names and channel topics and have them presented in a list window. You can also filter away unwanted channel names or topics and hide them from the view of employees or children. This feature can be password-protected to prevent the filters from being changed.

- *Channel folders.* You can store frequently joined or favorite channel names in a folder for future reference.

- *Chat schedule.* Launches a Web browser and connects you to the Chat Schedule page maintained on the World Wide Web by the vendor (Fig. 1.12) or a third party.

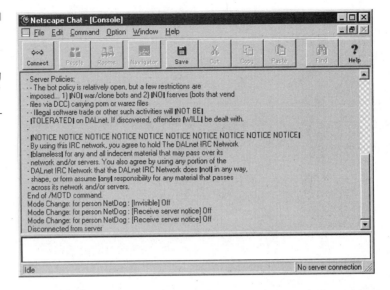

Figure 1.11
Netscape Chat offers
a simple way to get
started chatting using
the familiar Navigator
interface.

■ *Client-to-client protocol.* A protocol that allows you to retrieve information about a user, such as the person's real name, e-mail address, and information about the chat software in use. Chat software that supports this feature usually allows the owner of this information to suspend CTCP (client-to-client protocol) if frequent requests become

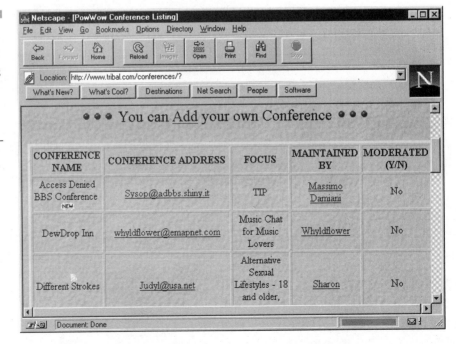

Figure 1.12
PowWow's Web
page allows you to
see what conferences
are scheduled, their
address, topic, main-
tainer, and whether
they are moderated.

annoying. With some software, users can even customize a response to such requests.

■ *Colored text.* Offers control of text color to ease reading of text conversations and differentiating them from server and control messages, which can be assigned their own colors.

■ *Direct client to client.* A feature that allows two parties to chat with each other directly, rather than through the IRC network. DCC also allows files to be sent and received between users. Users have the choice of whether to accept or reject such transactions.

■ *DNS lookup.* A feature that allows you to look up a person's domain name server. The information returned is the domain name and the IP address of the name server.

■ *Encryption.* Safeguards the privacy of chat sessions by applying encryption techniques to prevent text from being intercepted and read by a third party.

■ *Event handlers.* Allows you to write scripts that automate your response to specific types of events, such as people joining and leaving the channel you are on or your response to specific types of messages you receive.

■ *Finger.* This feature allows you to specify a user's address and receive back such information as login name, directory, date of last login, when new mail was received, and the last time that person read mail (Fig. 1.13).

■ *Font selection.* Offers control over the fonts used in the various windows of the chat program. You can specify overall font settings for all windows of the same type or set special fonts in specific windows.

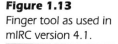

Figure 1.13
Finger tool as used in
mIRC version 4.1.

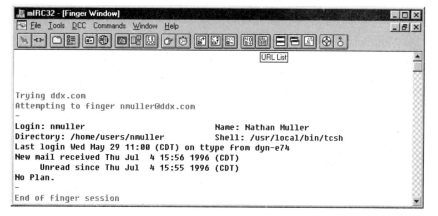

- *Freeze scrolling.*   Allows you to momentarily freeze the chat display so you can scroll back to follow a previous conversation.

- *Hotlist.*   Allows you to save the configurations of your favorite sessions so you can quickly reconnect to them in the future.

- *Macro functions.*   Allows you to create macros that are accessible via hot keys and/or a button bar.

- *MUD interface.*   Allows easy access to Multi-User Dungeons and provides the most common MUD commands and a compass for easy navigation.

- *Multiple session support.*   Allows you to establish multiple connections and switch between them.

- *Paste URL.*   Allows you to copy a currently opened URL from a Web browser and paste it directly into a chat session. This makes it easy to give guided tours of Web sites or share URLs with others online.

- *Personal biography.*   Allows you to compose a personal biography of yourself to share with other people you meet over IRC. The biography can include text and graphics in the same file.

- *Sound support.*   Allows you to send a beep or play an audio file to liven up your conversations in much the same way as the /ME command is used. Depending on the vendor, this feature can be used with or without accompanying text.

- *Split-screen support.*   Allows you to enter text in a separate window, so you can type without interruption as you continue to receive data from open connections.

- *Toolbar.*   Provides easy access to the most frequently used IRC commands.

- *URL launching.*   Allows you to click on any URL that appears on the chat screen so you can go to that location on the Web.

- *Voice editor.*   Allows you to modify the voice you want people to hear when they are speaking with you.

- *Web support.*   Provides easy access to the Web pages you talk about on IRC by supporting the major browsers, including Netscape, Mosaic, and MS Internet Explorer.

- *Whiteboard.*   A feature that allows you to share drawings with the people you chat with. It includes drawing, text, and color tools.

- *Yellow pages.*   Some chat programs include yellow pages to help you quickly find and connect to chat sites that fit your interests.

Browser Plug-Ins

Some chat software can be plugged into Web browsers such as Netscape Navigator. This allows visitors to interact directly with one another through a Web page without having to open a separate application to enter chat rooms. This integration offers visitors the best of both worlds: the ability to view Web pages while engaging in discussions with others viewing the same Web page. The addition of Web-integrated chat enables Web site operators—individuals and organizations—to develop a sense of community among their visitors who already have a common interest in the content of the site. It also increases visitor retention of the Web site.

Once installed in Netscape Navigator, Ichat will automatically launch when visitors connect to a Web site that is running the Ichat Interactive Server. Users see a Web page on the top two-thirds of their Navigator window, which is linked to an Ichat server taking up the bottom third of the Navigator window. The moderator of a room can lead guided tours of Web sites, prompting new pages to appear on a viewer's screen.

This server software extends that site's capabilities by adding real-time chat functionality and the concept of virtual space. The Ichat Interactive Server can be used to create virtually any type of custom object, including rooms and objects. Visitors who log in to the Ichat virtual world become objects, enabling them to move through and interact with other objects, as well as view other Web pages. The Ichat plug-in also allows users to click on hyperlinks to IRC and MUD networks.

Security

An unfortunate aspect of the Internet's skyrocketing growth is the increasing number of users who apply the technology for illicit and nefarious purposes. If you are concerned about the security of your conversations with others, at least one company, ComputerLink Online, offers a chat program with built-in encryption. The company's Secure Communicator for Microsoft Windows 95 uses proprietary encryption technologies it claims rivals those employed by Netscape Communications in its Web server product line.

Secure Communicator protects the chat session by encrypting all data sent between end points, locking out unwanted intruders without the need for overly complex firewalls or expensive security systems. The product can be used by businesses who are concerned about the security of their Internet commerce applications that include chat capabilities, as well as by individuals who require privacy for a wide range of personal reasons.

Online Resources

The following table provides the Web links of the major chat software vendors from which you can download working or evaluation copies of their software. The Web pages also contain such information as platforms supported, system requirements, product features, installation instructions, and troubleshooting advice. Since the technology is moving rapidly, you may want to consult these Web pages periodically for the latest developments.

Developer	Product	Web page or FTP site
Chris Bergmann	MacIRC	http://www.macirc.com/
ComputerLink Online	Secure Communicator	http://www.idirect.com/secure/
ELF Communications	WinTalk	http://www.elf.com/elf/wintalk/wintalk.html
Alex Fung	Winsock IRC	http://www.hkstar.com/~scfung
HiJiNX	InterFACE	http://www.hijinx.com.au
Ichat	Ichat	http://www.ichat.com
Internet Direct	Internet TeleCafe	http://www.telecafe.com/telecafe/
Khaled Mardam-Bey	mIRC	http://www-2.nijenrode.nl/software/mirc
MeGALiTH	Visual IRC	http://apollo3.com/~acable/virc.html
Netscape Communications	Netscape Chat	http://wuarchive.wustl.edu/packages/www/Netscape/chat/windows/
Northwest Computer Services	PIRCH	http://www.bcpl.lib.md.us/~frappa/pirch.html/
The Palace	The Palace	http://www.thepalace.com
Quarterdeck Corp.	Global Chat	http://www.qdeck.com/chat/
Tribal Voice	PowWow	http://www.tribal.com/
Worlds	Worlds Chat	http://www.worlds.net
George Xie	Xtalk	http://www.ugrad.cs.ubc.ca/spider/q7fl92/branch/tools.html
Z-Soft	Zchat	http://apollo3.com/~acowan/zchat.html

NOTE: This information, as well as updates, can be found at http://www.ddx.com/mgh.shtml.

Conclusion

When you first log on to a chat server, you might notice that many of the channels are sex-oriented and the names of the channels (and even people's nicknames) often use the most vulgar language. This shouldn't deter you from experimenting with and using chat software, however. If you look at the list showing the number of users participating on these channels, you'll see that the vast majority of these so-called adult channels are occupied by only one person—the one who created the channel in the first place. Enough said.

As the Internet evolves to better handle multimedia traffic, chat software will also evolve. Already, several chat software products have audio, voice, and video capabilities, including whiteboarding. In the future, we may see several types of products converge with very little to distinguish a chat product from a telephony product, for example. In fact, all of these capabilities are already incorporated in collaborative computing products to one extent or another. No matter how tightly these capabilities become integrated into specific products, vendors will always excel at one aspect, such as text chat or telephony, depending on how their products were designed to be used in the first place. Speaking of telephony over the Internet, that's the subject of the next chapter.

Telephone Calling Over the Internet

Introduction

One of the most exciting new applications of the Internet is telephone calling. With special phone software available from more than a dozen vendors, including IBM, Intel, Microsoft, and Netscape Communications, you can call virtually anyone, anywhere, anytime, and talk for as long as you want—free of local telephone company or long-distance carrier charges. If you pay a flat monthly fee for Internet access, these calls are absolutely free; if you pay an hourly charge for Internet access, at least the phone calls will cost no more than the normal hourly rate to use the Web. All you need is an account with an Internet service provider and some inexpensive hardware and software for your PC.

With the right combination of hardware and software—plus some fine-tuning—you can experience nearly the same voice quality as you would on a regular telephone call. The only difference is that your call is being carried over the Internet, rather than by a local telephone company or long-distance carrier. Depending on the phone software, or *phoneware*, you can even enjoy advanced call-handling features such as call waiting, call transfer, call conferencing, and voice mail—again, without having to pay extra monthly charges to a local telephone company or long-distance carrier.

Placing calls over the Internet is as easy as clicking on a name in the public white pages or your private phone book. *White pages* are simply public directories that are located at one or more servers on the Internet, while a *phone book* is a private list of names and addresses that resides on your own PC.

The public directories are maintained by the various phoneware vendors and are organized by user name and topic of interest, making it easy for users to strike up a conversation with like-minded and willing participants. Depending on vendor, directories are posted on special servers while others are posted on Web pages. In most cases, when you open the phoneware on your computer, it automatically registers you with a directory server (Fig. 2.1). Other callers can find you by viewing the directory, which is periodically updated and sent to your computer, reflecting changes as people enter and leave the network.

Several phoneware products even allow you to set up private directories on their server which can only be viewed by friends, family, or coworkers. Other phoneware vendors do not require that you register with a server before using their product. Instead, you can call someone merely by entering their IP (Internet Protocol) or e-mail address.

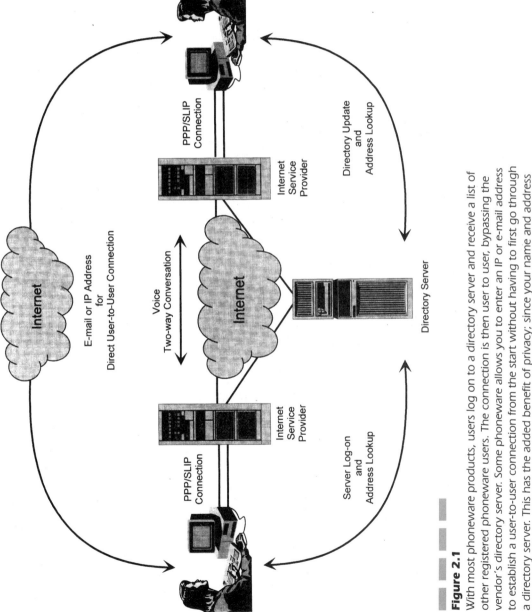

Figure 2.1

With most phoneware products, users log on to a directory server and receive a list of other registered phoneware users. The connection is then user to user, bypassing the vendor's directory server. Some phoneware allows you to enter an IP or e-mail address to establish a user-to-user connection from the start without having to first go through a directory server. This has the added benefit of privacy; since your name and address do not show up on a public directory, you will not be bothered by unwanted calls.

Hardware Components

To use phoneware, you must have a computer with enough processing power to handle the voice compression/decompression algorithms the phoneware vendors use. In essence, these algorithms package samples of your voice into packets suitable for transmission over the Internet. At the other end, the packets are opened and your sampled voice is rebuilt, preserving most of its unique characteristics. For this to work properly, you need a 486/33DX computer or better. With anything less—even a 486 SX—there will be too much processing delay, causing problems that range from lost syllables to total operational failure.

You will also need enough memory, between 4 and 8 MB, depending on the product you choose. Although some phoneware vendors claim you only need 4 MB of RAM to run their product, you will probably be running other things at the same time. For instance, if you run Windows 95 and Netscape Navigator at the same time you run the phoneware, you'll need at least 8 MB of RAM. To provide an extra margin of safety, 16 MB of RAM is recommended.

Phoneware products typically do not take up any more hard disk space than other graphically enriched applications. Depending on the product, you'll need only 1 to 5 MB of available disk space for a full installation—more, if the phoneware product comes with optional add-ons or is bundled with a Web browser.

The right modem is essential for any serious Internet user and more so if you expect to realize the full potential of the phoneware. At a minimum, a 14.4 Kbps modem is required to move packets of voice fast enough over the Internet with acceptable delay. A 28.8 Kbps modem reduces the delay even more, giving phone conversations a much smoother flow. If you're still using a 9.6 Kbps modem, it is best not to buy any other hardware or software until you upgrade your modem.

To use phoneware, you'll also need a Windows-compatible sound card, speakers (or headset), and a microphone. If you are installing a sound card for the first time, it is recommended that you test it for proper operation before installing the phoneware. That way, if you experience any problems you have only one possible source—the phoneware—to worry about instead of two. Some sound cards have a setup utility that senses what resources are already in use and automatically configures the board to avoid most conflicts. If the test feature detects a conflict between various COM port and IRQ combinations, for example, an appropriate notification is displayed so you can select another setting.

There are two types of sound cards: half duplex and full duplex. *Half-duplex* mode works like a citizen's band (CB) radio, where one person can talk at a time and says "over," indicating when he or she is finished talking. This method of communicating requires patience and there's always the chance of the other party hogging the connection, preventing you from getting in a choice word or two. With *full-duplex* audio cards, both parties can talk at once, just like an ordinary telephone call. However, if a full-duplex user is connected to a half-duplex user, the conversation defaults to the half-duplex mode.

A common problem with full-duplex conversation is acoustic feedback. But this can be minimized by using a headset to better isolate the received signal from the microphone. Alternatively, you can lower the speaker volume and put some distance between the microphone and speaker.

An economical alternative to computer speakers and microphone is to use a telephone handset that plugs directly into sound cards that use standard ⅛-inch jacks. One product, NET-FONE, costs less than $40, does not require any external power source, and has no drivers that need to be installed. The drawback is that you must hold the handset to your ear. For hands-free operation, you can get an earpiece from JABRA Corporation called JABRA Ear PHONE. The unit consists of a lightweight, single-piece microphone and speaker that plugs into your ear. The unit costs about $90. Of course, there are less exotic microphone/speaker alternatives available via mail order which cost under $10.

You can get a good deal on bundled products. For example, you can buy a sound card that includes a headset with built-in speaker and a microphone from mail-order firms for as low as $125.

Software Components

To use the Internet for phone calls, you will need three software components. Two of them you probably already have: a TCP/IP dialer program and a Web browser. The other is the phone software itself. If you are concerned about the security of your voice conversations, you have a choice of using optional encryption software or using a phoneware product with built-in encryption.

TCP/IP Dialer

As noted in Chap. 1, modem access to the Internet is accomplished with a TCP/IP dialer program, which enables information to be transmitted in packets. If you already have an Internet access account, you're probably already using this software. Trumpet and TwinSock are two popular shareware programs that are known for their ease of use. Another easy-to-use TCP/IP dialer program comes with Windows 95. To use the Windows 95 dialer program you must verify that the dial-up adapter and TCP/IP protocol are installed. Just press the Start button, go to Control Panel, and double-click the Network icon. With the Configuration tab selected, both Dial-Up Adapter and TCP/IP should be listed. If not listed, these items must be installed. If already installed, these two components must be properly configured with information given to you by your Internet service provider.

Phoneware

The obvious software you will need is the phoneware itself. Assuming that you have a reasonably powerful PC and that it is equipped with the appropriate hardware, you are probably running Windows 3.1 or have upgraded to Windows 95. There are commercial phoneware products for each of these, in addition to products for the Macintosh, such as Maven, PGPfone, and ClearPhone. There is even phoneware for UNIX systems and phoneware that work across multiple hardware platforms.

One of the oldest phoneware products is vat, which traditionally has been a UNIX program. A new version has recently become available for Windows 95 and Windows NT. The latest version of Maven, which runs on Macintosh computers, is interoperable with vat running on UNIX machines. Speak Freely offers cross-platform communication between Windows and UNIX users, while CU-SeeMe and VocalTec's Internet Phone are among the phoneware products that offer cross-platform communication between Macintosh and PC users.

Basically, the phoneware provides algorithms that compress the recorded speech obtained from the sound card and employs optimization techniques to ensure its efficient delivery over the Internet. Phoneware vendors use a variety of compression algorithms to minimize bandwidth consumption over the Internet. For example, NetSpeak's WebPhone uses two audio compression algorithms—GSM and TrueSpeech. GSM is the Global System for Mobile communications and is a worldwide standard for digital cellular communications. It provides close to a 5:1 compression of raw audio with an acceptable loss of audio quality on decompression. TrueSpeech, a product of the DSP Group, provides compression ratios as high as 18:1 with an imperceptible loss of audio quality on decompression. WebPhone uses

GSM compression when it is installed on a 486-based computer and True-Speech when it is installed on a Pentium-based computer. Offering a high compression ratio, TrueSpeech is more CPU-intensive than GSM, so it requires a faster processor to compress the same audio signal in real time.

VocalTec supports three compression schemes in its Internet Phone product: TrueSpeech, GSM, and its own VSC coder which squeezes raw audio data down to 7.7 Kbps. Users have the option of adding VocalTec's VC Card to reduce bandwidth consumption further, to about 6.72 Kbps, which provides the added benefit of offloading this chore from the computer's main processor. Although the VSC scheme is not as efficient as GSM, it is designed so that it can be used on a variety of hardware platforms, regardless of processor type.

Netscape's CoolTalk also supports GSM, which would be used for general purpose conferencing. But under low-bandwidth conditions for speech-only conferencing, it uses a compression algorithm called RT24. The choice is user-selectable.

You should try out various phoneware products before you buy. They differ in terms of installation ease, features, and performance. Most vendors have a Web page or FTP site from which trial versions of the software can be downloaded for evaluation. During installation, you will have the opportunity to register your name with the directory server (if required) and specify any discussion groups you would like to belong to. Depending on vendor, you may not have access to all the advanced features and you might even be limited to only a few minutes of conversation with each call until you buy the software. In most cases, this can be done conveniently over the Internet via an order form posted on the vendor's Web page, provided that you want to supply a credit card number.

For example, a limited version of NetSpeak's WebPhone (Fig. 2.2) is available for download at http://www.netspeak.com. This "unactivated" version has the following restrictions:

■ Three minutes of talk time per call versus unlimited talk time with the activated version.

■ One line may be used versus use of four lines simultaneously with the activated version.

■ A maximum of three phone directory entries are allowed versus an unlimited number of directory entries with the activated version.

■ Two received voice mail messages can be retained versus an unlimited number of voice mail messages with the activated version.

■ One custom outgoing message may be defined versus an unlimited number of messages with the activated version.

Figure 2.2
(*a*) NetSpeak's Web-
Phone interface looks
like a cellular flip-
phone keypad.

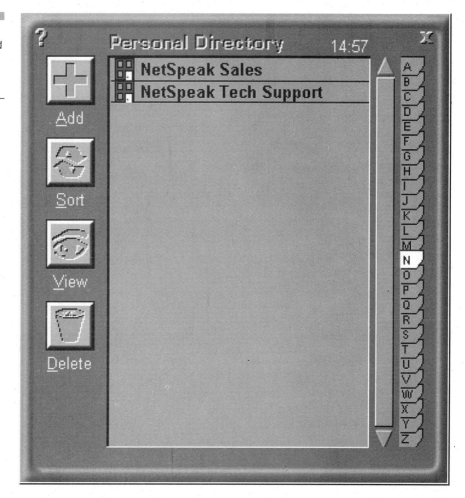

Figure 2.2
(b) The interface used for the address book is familiar and it is friendly.

These restrictions are removed upon payment of a one-time fee, which also entitles the user to free product upgrades.

There are a few products that allow unlimited talk time during the evaluation period, such as SilverSoft's SoftFone. One commercial product, FreeTel, involves absolutely no cost to the user. The provider, FreeTel Communications, seeks to support itself with advertising that operates like an electronic billboard on the client screen (Fig. 2.3). The advertising is downloaded in the background, while you are waiting to make a connection. You can even click on advertiser links that strike your interest.

Be aware that the different phoneware products do not interoperate with each other. If you use VocalTec's Internet Phone, for example, anyone you wish to communicate with must also use VocalTec's Internet Phone. You will not be able to communicate with users of Camelot's DigiPhone or

Figure 2.3

FreeTel Communications offers its phoneware at no cost. It is entirely supported by vendor advertising which appears at the top of the user interface. In addition to voice, FreeTel provides a split screen for text chatting, which is a handy feature to have if your connection deteriorates during the voice conversation or if you want to spell out a difficult name, word, or phrase.

NetSpeak's WebPhone or any other competing product. This means that if you plan to use these products to communicate with family and friends or business colleagues, some coordination is needed—before everyone decides to purchase on their own and discover later that they can't talk to each other! This situation is already changing, as discussed later.

Web Browser

The only other software you will need is a Web browser such as Mosaic or Netscape Navigator—or some other graphical Web navigation software. Some phoneware products do not require that the browser software be opened to make calls over the Internet. If the browser is required, usually it must be opened up before you attempt to use the phoneware.

Security

If you're looking to protect your conversations as they traverse the Internet, there are at least three products that offer encryption that turns your computer into a secure telephone: MIT's PGPfone, Camelot's DigiPhone, and NetSpeak's WebPhone.

As its name implies, PGPfone uses a cryptography technique called PGP (Pretty Good Privacy) to enable real-time secure telephone conversation. PGPfone is available in Windows 95 and Macintosh versions. PGP generates two keys that belong uniquely to you. One PGP key is secret and stays in your computer. The other key is public and is given out to people you want to communicate with.

PGP is a separate program that is used by PGPfone. PGP is distributed by MIT with the understanding that you are a U.S. citizen located in the United States or a Canadian citizen located in Canada. The PGP Distribution Authorization Form is located on the Web at

http://web.mit.edu/network/pgp-form.html

NetSpeak's WebPhone also offers encryption of conversations. The advantage of using phoneware products with integral encryption is that they are easier to use than if you must configure encryption programs separately. Another phoneware product that employs encryption is Camelot's DigiPhone. Both products use alternative voice encryption techniques rather than PGP to safeguard the privacy of Internet calls.

The Internet Connection

The speed of the Internet is becoming more of a concern for everyone, especially users and content providers on the World Wide Web. As usage increases, so does the likelihood of delays in establishing the access connection, and once connected, in doing lookups and obtaining information on the Web in a timely manner.

In addition to the algorithms that compress/decompress sampled voice, phoneware products may include optimization techniques to deal with the inherent delay of the Internet. The packets may take different paths to their destination and may not all arrive in time to be reassembled in the proper sequence. If this was ordinary data, late or bad packets would simply be dropped and the host's error-checking protocols would request a retransmission of those packets. But this concept cannot be applied to packets containing compressed audio without causing major disruption to voice conversations, which are supposed to be conducted in real time. If only a small percentage of the packets are dropped, say 2 to 5 percent, the users at each end may not notice the gaps in their conversation. When packet loss approaches 20 percent, however, the quality of the conversation begins to deteriorate. Some products, such as VocalTec's Internet Phone, employ predictive analysis techniques to reconstruct lost packets, thereby minimizing this problem.

Occasionally, the Internet or your ISP can become overloaded or congested, resulting in lost packets and choppy sound quality. FreeTel has a feature called Booster that alleviates this problem and improves sound quality. You can enable or disable the Booster during a conversation as conditions warrant. Booster introduces an artificial delay into the signal, and uses this extra time to retransmit lost packets using a proprietary algorithm. The end result is better quality, at the expense of increased but predictable delay. Many people use this feature for overseas connections where the loss rate can be significant.

Putting It All Together

There are two aspects to preparing your PC for Internet phone calls: configuring the hardware and configuring the software. You will have to make the connections between the modem and sound cards and make sure all the components (i.e., microphone and speakers or headset) are properly connected. Then you will have to install and configure the phoneware.

Configuring the Hardware

If you have a multifunction modem/sound card and it has been properly installed in your PC per the manufacturer's instructions, the next step is to make sure all the cables get connected to the right places before you turn on the PC (Fig. 2.4).

To start with, the microphone gets plugged into the modem/sound card at a port that should be labeled as such—usually *MIC.* You can stick the microphone to the top of your computer's monitor or simply place it on the desk next to your keyboard.

The speakers (or headset) plug into the *line out* port on the modem/sound card. (The *line in* port is a connection to your CD player, but this is not necessary for the operation of phoneware.) Speakers should be positioned well away from the microphone to prevent *feedback,* a situation that results when the microphone picks up sound from the speakers.

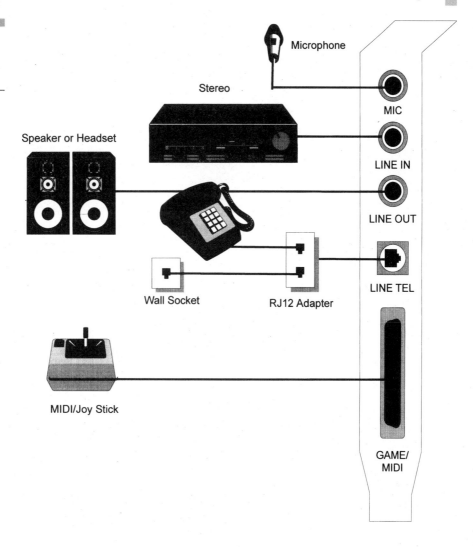

Figure 2.4
Typical connections
to the multifunction
modem/sound card.

When this happens, the user at the other end hears his or her own voice as an echo. If feedback is still a problem, you can adjust the sensitivity of the microphone until it stops or avoid the problem entirely by using a headset.

The modem/sound card has a modular cable port, which is usually labeled *line tel*. This is where you plug in the RJ12 adapter. One port of the RJ12 adapter is used to plug in the phone and the other is used for the connection to the wall socket. This will allow you to use your telephone when you're not connected to the Internet. The fax/modem card will have a MIC port for a microphone. This microphone is used to record

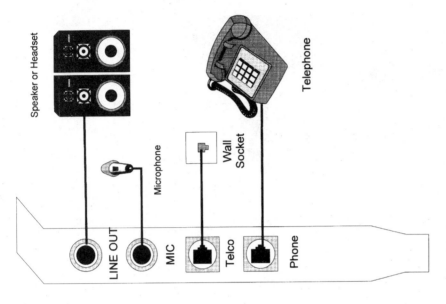

Fax/Modem Card

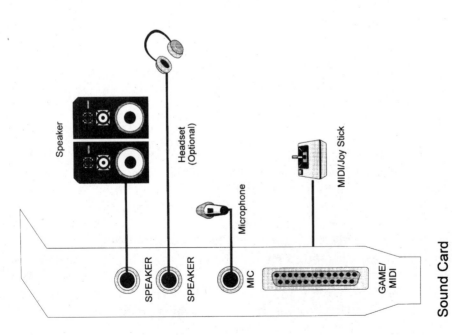

Sound Card

Figure 2.5

Typical connections to a fax/modem card and sound card.

greeting and menu messages when the fax/modem card is configured to automatically answer phone calls. It is not used with the phoneware.

If you have a separate fax/modem card (or just a modem card) and sound card, the connections are slightly different, as illustrated in Fig. 2.5.

You plug in your phone cable to the phone port and use the line port for a connection to the wall jack. You connect the microphone to the sound card's MIC port. This is the microphone you will use with your phoneware. The modem and sound card work together via the PC's internal bus connections, so no external cabling is necessary to link the two devices.

Since every modem, fax/modem, and sound card is configured differently in terms of the types of ports, the arrangement of ports, and labeling of ports, it is recommended that you read the product documentation for specific cabling and configuration instructions.

Configuring the Phoneware

After you install your phoneware according to the vendor's instructions, you will typically need to enter your communications parameters, such as communications port number and IRQ. You will also need to enter some basic Internet-related information, the most important being:

- *E-mail address.* Such as *nmuller@ddx.com*
- *IP address.* Such as *204.177.193.70*
- *POP Server address.* Such as *vespucci.iquest.com*
- *SMTP Server address.* Such as *vespucci.iquest.com*
- *E-mail login and password.* Such as *nmuller and ********

The POP server address and SMTP server address are probably the same. If you intend to operate your phoneware on a LAN- or WAN-connected computer, your network probably does not have an SMTP or POP Server. In this case, you would leave everything but the IP address field blank. Alternatively, if you use a dial-up connection to your Internet service provider, you probably do not have a fixed IP address and should leave that field blank. This point deserves clarification.

There are two types of IP addresses: dynamic and static. As its name implies, a *dynamic* IP address is assigned to you by your Internet service provider each time you dial in to the server. The Internet service provider has a pool of IP addresses for this purpose. With *static* IP addressing, your machine uses the same IP address each time you connect to the Internet.

The proliferation of TCP/IP-based networks, coupled with the growing demand for Internet addresses, made it necessary to conserve IP addresses. Issuing IP addresses on a dynamic basis provides a way to recycle this finite resource. Even companies with private intranets are increasingly using dynamic IP addresses, instead of issuing unique IP addresses to every machine. There is even a new standard that addresses this issue—the *Dynamic Host Configuration Protocol* (DHCP), developed by the Internet Engineering Task Force (IETF). From a pool of IP addresses, a DHCP server doles them out to users as they establish Internet connections. When they log off the Net, the IP addresses become available to other users.

With the phoneware software installed, it is time to get on the Internet and start talking. With your modem turned on, open the Winsock application to establish the Internet connection. When you get an indication that packet mode has been enabled, start up your Web browser or e-mail program—if the phoneware product you have requires it. If not, you can click on the phoneware application to start it up. The next step is to establish a connection. You can do this by connecting to the directory server or by entering a specific IP address.

Establishing a connection to the directory server is as easy as clicking on a log-on button. Once you register with the server, you can pick names off your private directory or select names off the public directory to initiate calls. Or you can wait for someone to call you upon seeing your name displayed in a directory.

Alternatively, you can bypass the directory server entirely by entering the IP address of the person you want to call. This will only work if that person has a permanent (i.e., static) IP address and is on the Internet the same time you are. As noted, if the user is on a corporate network connected to the Internet, there may be firewalls or proxy servers standing in the way to implement security policies. In such cases, your call might not get through.

If you want to check whether a call will get through to a particular IP address before actually making the call, you can use a ping utility (Fig. 2.6). A *ping* is a network packet which is sent from one computer to another and then echoed back for the purpose of measuring packet delay (in milliseconds) and packet loss between two hosts on the network. This is done by using a ping utility that sends a signal to the remote machine which essentially asks, "Are you there?" If the remote machine is turned on and is able to send and receive messages, it will respond to the ping's inquiry. If there is no response, this means the remote machine is not functioning or a firewall has blocked the ping

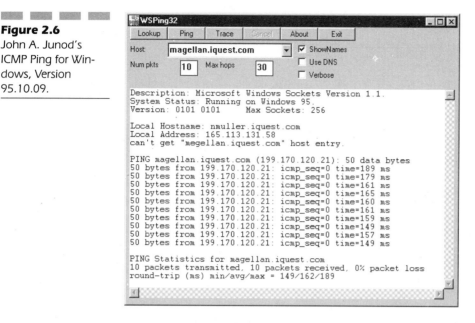

Figure 2.6
John A. Junod's
ICMP Ping for Win-
dows, Version
95.10.09.

signal. In either case, you will not be able to communicate with the desired IP address.

As you speak into the sound card's microphone, the Internet phone software samples the incoming audio signal, compresses it, and transmits the packets via TCP/IP over your communication link to the remote party. At the other end, the packets of compressed audio are received and pieced together in the right order. The audio is then decompressed and sent to the sound card's speaker for the other party to hear. Much of the Internet's inherent delay is compensated for by the compression algorithm. As the packets are decompressed and the signals are being played, more compressed packets are arriving. This process approximates real-time conversation.

You can run the phone software in the background with its window minimized, just like fax software, allowing you to work in other applications. A notification tone or visual indicator lets you know when a call is coming through. Clicking on the phone application's icon opens the user interface where the caller's name or nickname is displayed. You can then choose to answer or ignore the call. If you do not happen to be running the phone software when a call comes through, the calling party will be notified that you are not connected. If you're already engaged in a conversation when a call comes through—or you don't want to accept the call—the calling party will get a busy notification. With one product,

Camelot's DigiPhone, you have the option of sending an e-mail message to a busy or unconnected party, letting him or her know you tried to call.

```
Date: 30 Sep 96 19:22:15
From: lmuller@iquest.com
To: nmuller@ddx.com
cc:
Subject: DigiPhoned while you were out....

You were unavailable when I called. Please call me as soon as is
convenient.
```

And with VocalTec's Internet Phone, on-screen messages let you know if someone is busy or not connected when you tried to call. If you were too busy to take a call, a message is displayed letting you know who called.

If you are concerned about receiving junk calls on the Internet, you may want to try TS Intercom from Telescape Communications (Fig. 2.7). This phoneware product is specifically designed for private communications. This means your name and e-mail address are not published on an Internet server and no one will try to call you while you are online, as is the case with products modeled after public IRC chat channels. Instead, you connect to people directly by entering their e-mail address. In this way, privacy is assured. The company does offer the option of allowing

Figure 2.7
By simply entering the e-mail address of the person you wish to speak with, TS Intercom from Telescape Communications provides direct user-to-user voice communication over the Internet. Other phoneware products offer this capability as well.

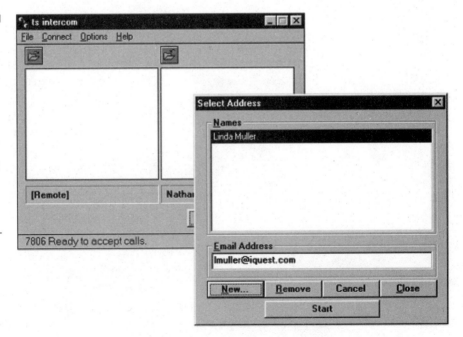

you to publish your name in a user database to help others locate you. The database is organized by country and areas of interest. The database is even available for your use without having to first add your personal information and it can be downloaded to your computer in its entirety.

Through a Viewport in the Windows graphical user interface, you can display a visual signature, photograph, or any graphic file to whomever you are talking. The software itself is freely distributable, so you can send a copy to whomever you want to talk and start conversing over the Internet immediately. Since calls are initiated with only an e-mail address, the need for product registration codes is eliminated, allowing you to use the software as soon as you install it on your computer.

Fine-Tuning

Fine-tuning the sampling rate and compression level to your modem's speed will improve overall sound quality. While the average delay on the Internet is generally not a serious problem, frequent instances of clipped speech can be quite annoying. If this occurs, you can reduce the sampling rate until smooth speech resumes.

With DigiPhone, for instance, you can dynamically adjust its recording and playback quality in response to the speed of your modem connection. You can start your manual tuning process by connecting at the default sampling rate, then increase the sampling rate by increments of 500—the increments can be smaller or larger depending on your preference. When you're finished, DigiPhone will renegotiate the connection at the higher sampling rate. You can continue to increment the sampling rate until the other party's speech begins to break up. Then back down until it is clear again.

The sampling rate can be set from 4000 to 44,000 bytes per second, depending on the capabilities of your sound card. In general, the higher the speed of your modem connection, the higher you can set the sampling rate. Here are some starting points:

Connection Speed	Sampling Rate
9,600 bps	4,000
14,400 bps	6,000
19,200 bps	8,000
28,800 bps	11,000

If you are using a lower-speed modem, you can select a lower sampling rate to achieve better performance. However, better performance will come at the expense of sound quality, which will be lower. If you have a higher-speed modem, you can select a higher sampling rate to achieve better overall sound quality. In conjunction with the sampling rate, you can set the compression level. With a lower-speed modem, you can select a higher compression level for better performance, but with some loss in sound quality. With a higher-speed modem, you can select a lower compression level for better sound quality.

VocalTec's Internet Phone provides real-time statistics that can help you determine the quality of the Internet connection at any given time (Fig. 2.8). The network statistics window provides a count of incoming and outgoing packets, the average round-trip delay of packets, lost packets on the way to your system, and lost packets on the way to the other person's system. The window also displays *send errors,* which is the number of packets that were not sent due to a TCP/IP, modem, or CPU problem.

A common problem with using phoneware is that of being able to hear others and having them hear you. This happens when the sound card ships with the microphone input disabled or the sound level turned off. Most sound cards come with a set of utilities, including

Figure 2.8
Network statistics window of VocalTec's Internet Phone.

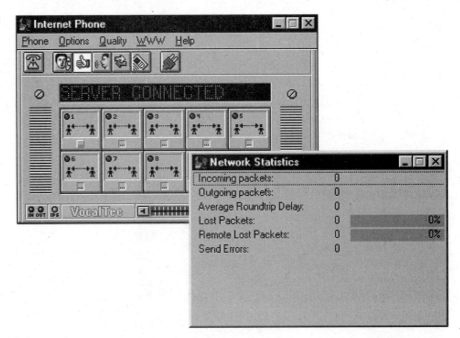

a mixer program. With the mixer utility, you can enable the card's microphone input and adjust the volume. Depending on the type of sound card you have, it is often possible to adjust microphone volume from the phoneware's interface. With Quarterdeck's WebTalk, for example, there are three microphone settings on the main screen (Fig. 2.9), as follows:

- *Sensitivity.* Controls microphone volume
- *Idle level.* Controls how loudly you must speak to activate the microphone which helps prevent background noise from being picked up while you speak
- *Idle after.* Controls how soon after you stop speaking before control of the microphone is turned over to the other user

Some sound cards provide *Automatic Gain Control* (AGC). This feature boosts the microphone level automatically when you're speaking into it and reduces the level when you're not speaking. This cuts down on ambient background noise that makes it difficult for both parties to hear each other. You must make sure you have AGC turned on. You can then adjust it by accessing the sound card's configuration utility which offers a slider control for adjusting bass and treble.

Figure 2.9
Microphone settings provided by Quarterdeck's WebTalk.

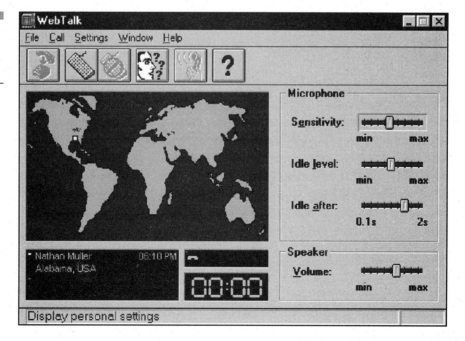

Features

Internet phone products differ in terms of performance and features. After narrowing down your choices to a few products based on how well the vendors' evaluation software performed, you can turn your attention to comparing features to make a final selection. Here is a list of features you may want to look for:

- *Adjustable volume control.* Allows you to adjust the volume of the microphone and speakers during the conversation.

- *Advanced caller ID.* Not only is the calling party identified by name, but some phoneware products offer a brief introduction message about what callers want to talk about, which is displayed as the call comes through. You can use this information to decide whether or not to answer the call.

- *Advanced phone book.* Not only holds contact information, but offers a search capability by name, e-mail, country, company, or any other parameter that can identify a particular person.

- *Audio date/time stamp.* Informs you when callers attempted to reach you by date and time.

- *Automatic notification.* With this feature, the phoneware automatically looks for and notifies you when users come online so you can call them.

- *Busy notification.* If you place a call to someone who is busy with another call, you will get back an appropriate message. Some products allow you to send an e-mail message, voice mail, or other notification to the busy party indicating that you tried to call.

- *Call blocking.* You can choose to block annoying or unwanted incoming calls by fixed IP or e-mail address.

- *Call conferencing.* The ability to converse with three or more people at the same time. (Same as *multicasting*.)

- *Call duration timer.* Provides an indication of the amount of time spent on each call.

- *Call hold.* Allows you to put the initial call on hold while you answer another incoming call. You may continue the first conversation after holding or hanging up the second call. (See also the description of *music on hold* that follows.)

- *Call log.* Records information about incoming and outgoing calls, allowing you to keep track of who called you or who you called.

- *Call queue.* A place to put the person you wish to talk with. If that person is not available to take your call, the phoneware automatically redials every three minutes until the connection is established.

- *Caller ID.* Identifies the caller by name, nickname, and/or e-mail address so you can see who is calling before deciding whether to take the call.

- *Configuration utility.* Scans your system to determine if the proper hardware is installed to use the phoneware and offers advice on configuring various operating parameters such as IRQs, DMAs, and I/O base address settings to prevent conflicts with other communications applications.

- *Database repair utility.* The phoneware maintains one or more databases to hold such things as private phone books and configuration data. If a database gets corrupted or destroyed, you will be notified and have the option of running the database repair utility to restore it.

- *Dedicated server.* For those who receive a large volume of incoming calls, some phoneware vendors offer special servers to faciliate call handling.

- *Directory assistance.* A searchable directory of users currently online is automatically maintained. You can initiate a phone call simply by mouse-clicking on a person's name or by typing in the first few characters of a name.

- *Dynamic, on-screen directory.* Provides you with the latest information on users who have registered with the server, indicating that they are online and ready to take or initiate calls. This display is periodically refreshed with new information.

- *Encryption.* To ensure secure voice communication over the Internet, you can use a public-key encryption technology such as PGP (Pretty Good Privacy). Depending on vendor, PGP is integrated into the phoneware or may be licensed separately for use with the phoneware. However, other products do not accommodate encryption at all.

- *Event message system.* Allows you to view the ongoing status of the phoneware so you know what features and functions are active at any given moment.

- *File transfer.* You can transmit a file to the other party during a conversation. The file transfer process takes place in the background and will not interfere with the conversation.

- *Greeting message.* A recorded message that is played to callers when you are away from your PC or too busy to answer.

- *H.323 compliance.* A worldwide standard for audio and video communication over packet-switched networks, such as the Internet. Users of H.323-compliant products are capable of talking with each other over the Internet.

- *Last party redial.* Allows you to redial the last party called without having to look up the address in a directory.

- *Map.* Displays the connection of the call against the background of a U.S. or world map, depending on where you are calling.

- *Multicasting.* This allows voice conversations between three or more users on a selective basis. (Same as *call conferencing.*)

- *Multiple calling mechanisms.* Some phoneware products offer multiple methods of initiating calls, including by fixed IP address, domain name, e-mail address, saved addresses, and online directory of registered users.

- *Multiple lines.* Some phoneware products allow you to carry on a conversation on one line and take an incoming call on another line, or to put one call on hold while you initiate another call.

- *Multiple-user configurations.* If several people share the same computer, some phoneware products allow each of them to have their own private configuration, including caller ID information.

- *Music on hold.* Plays music to a caller on hold, until you are through with another call.

- *Mute.* A mute button allows private, offline conversations.

- *Online help.* Offers help on the proper use of various phoneware features without having to resort to a manual or opening a separate read-me file.

- *Picture compression.* Some phoneware allows the user to call up a photo of the person he or she is talking to (if the remote user supplies one). Compression allows fast photo loading over the Internet of the commonly supported file formats, including GIF, JPG, TGA, PCX, and PBM.

- *Programmable buttons.* Allows you to configure quick-dial buttons for the people you call most frequently. In some cases, buttons are added automatically and written over based on the most recent calls.

- *Remote information display.* Displays the operating system and sound card information of the remote user.

- *Remote time display.* Displays the remote time of the person you are talking to.

- *Selectable codec.* Provides a choice of codecs, depending on the processing power of the computer. A high-compression codec can be used for Pentium-based machines and a lower-compression codec can be used for 486-based machines. The choice is determined during phoneware installation.

- *Silence detection.* Detects periods of silence during the conversation to avoid unnecessary transmission.

- *Statistics window.* Allows you to monitor system performance and the quality of the Internet link.

- *Text chat.* Some phoneware products offer an interactive text or chat capability to augment voice conversations. In some cases, the chat feature can be used before actually answering in voice mode.

- *Toolbar.* Icons provide quick access to frequently performed tasks such as hang up, mute, chat, view settings, and help.

- *Toolbox mode.* The interface can be collapsed into a compact toolbox to save desktop space. This makes it easier to work in other applications until the phoneware is used for calls.

- *User Location Service (ULS) compliance.* ULS technology enables Internet phone users to find each other through existing Internet Directory Services such as Four11, Banyan's Switchboard, WhoWhere, DoubleClick, and BigFoot.

- *User-defined groups.* Allows users to set up private calling circles for calls among members only or establish new topic groups for public access.

- *Video.* Some phoneware products originated as videoconferencing products and have added interfaces that allow them to operate in audio mode for voice conversations only. Other products originated as text chat products and have added an audio capability. They provide the means to send a still image of each participant in GIF or JPEG format.

- *Voice mail.* Allows you to record and playback greeting messages as well as send voice mail messages for playback by recipients. Depending on vendor, this might include the ability to give specific messages to callers when they enter a personal code.

- *Voice mail screening.* Allows you to delete voice messages before they can be downloaded to your computer.

- *Web links.* You can put links to some phoneware products in your Web pages, allowing those using the same phoneware to call you simply by clicking on the link.

Figure 2.10
Camelot's DigiPhone
interface.

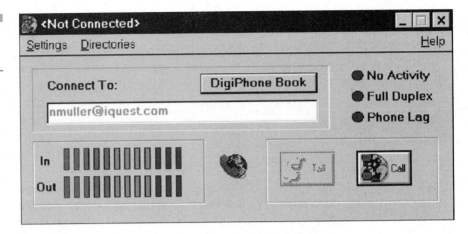

- *Whiteboard.* Some multifunction products allow participants to draw or annotate shared text and images while conversing.

Camelot allows purchasers of its DigiPhone (Fig. 2.10) to download a second license free from its Web page. Quarterdeck includes a copy of Mosaic and a free second license with the purchase of its WebTalk. Some vendors offer free upgrades for commercial phoneware products six months to a year after the purchase. In some cases, the vendor may even allow users to install the phoneware on multiple machines.

In the event you change your e-mail address, you will need to inform the phoneware provider so you can receive a new registration file or code—if one is required. Some vendors reserve the right to request proof of your e-mail address change.

Product selection should reflect your priorities and those of the people you want to converse with over the Internet. Don't forget—different phoneware products may not be interoperable, in which case you can communicate only with people who use the same phoneware you use. Product selection requires some coordination and you might have to compromise to arrive at a consensus.

Internet Calls to Conventional Phones

In an experiment to demonstrate the feasibility of originating calls on the Internet and receiving them at conventional phones in select locations around the world, the Free World Dialup (FWD) Global Server Network project went online in March 1996. Organized by hundreds of volunteers around the world, the noncommercial project was entirely coordinated in cyberspace via Internet telephony, e-mail, and chat software.

Using popular Internet telephony software such as VocalTec's Internet Phone or Camelot's DigiPhone, users contact a remote server in the destination city of their call. This server *patches* the Internet phone call to any phone number in the local exchange. This means, for example, that a user in Hong Kong can use an Internet-based server in Paris to effectively dial any local phone number and talk with a friend or family member for free. A global server keeps a list of all servers and the real-time status of each. The FWD system is regulated by a client-server structure and software that the project group has developed for use in different countries.

The specific steps required to place a call from the Internet to a conventional phone are as follows:

1. Connect to the Internet as you normally do with a PPP or SLIP connection.

2. Start the phoneware and register with the vendor's server, if necessary.

3. Start the FWD client software and connect to a FWD server in a select city.

4. Click on the Connect button and enter the domain name of the server.

5. Once connected, a message indicates that you are indeed connected. You then hear a ring as though an incoming call was coming through.

6. Answer the call and enter the telephone number in the FWD client of the person in the local calling area, leaving out the area code.

7. Click on the dial button in the FWD client software.

8. You should hear the phoneware dial the number you entered.

9. When the called person answers, both parties can start conversing.

The procedure might seem complicated at first, but it gets easier after dialing a few calls. Nevertheless, the procedure is cumbersome compared to dialing an 11-digit long-distance number. This is one reason why Internet telephony will not pose a threat to long-distance carriers for some time to come.

To become a server on the FWD system—and thus expand the geographical coverage of the service—participants will need one connection to the Internet (either dedicated or dial-up) and another phone line to patch calls to their local calling area for a total of two phone lines. Participants also need a Windows machine and sound card, as well as a voice modem which uses the Cirrus Logic chipset to patch through the calls. The FWD project supplies all the software (at no charge) needed to enable the computer to function as a server. At this writing, there are 45 community servers operating in over 40 countries and applications have been received by hundreds of individuals who want to set up servers to support the service. This indicates that the FWD network has the potential to evolve beyond a mere experimental project and grow at a fairly rapid pace.

Right now, call origination is one-way—from the Internet to conventional phones. In the future, a dial-in capability may be added. This would allow people to call into the FWD system and around the world via the Internet from any Touch-Tone telephone, creating a seamless system where users on both sides do not need to have a computer or Internet connection.

The FWD project has its own home page on the Web at

http://www.pulver.com/fwd/

The Web page provides updates on the progress of the project, including how to become a volunteer server to support expansion of the service. New server locations are also posted on this Web page.

Another Web page provides an online Web board that allows participants to ask questions and get technical assistance on using FWD. This page is located at

http://www.pulver.com/fwd/wwwboard/

The FWD project maintains a FAQ (frequently asked questions) page at

http://www.pulver.com/fwd/new/faq.htm

Independently of the FWD experiment several phoneware vendors have turned the concept into products that can be used to provide commercial calling services, among them: VocalTec and NetSpeak.

VocalTec's Internet Phone Telephony Gateway provides connections between the Internet and local telephone lines, enabling callers to combine the low cost of Internet connections, the convenience of initiating calls from either PCs or telephones, and the ability to communicate with anybody with a telephone via the public switched telephone network. Calls to wireline and cellular phones are possible, and the cost of a call is limited to the telephone charges incurred when linking to the Internet on either end of the connection, plus the standard Internet connectivity charges.

VocalTec's system is comprised of a PC running Windows 95 and the Internet Phone Telephony Gateway software, equipped with a Dialogic Corporation computer telephony card and linked to the telephone network and the Internet through a 28.8-Kbps or faster connection. The Internet Phone Telephony Gateway not only gives people with PCs a way to talk over the Internet with people who do not use a computer, it enables value-added resellers (VARs) and systems integrators to offer innovative applications such as international "hop off" and Internet-based customer service.

NetSpeak offers a client-server-based Business WebPhone System for companies that want to give employees the means to place telephone calls over the corporate intranet as well as the Internet. Callers can set up phone conferences between up to four people, leave voice mail, and forward calls. The server runs on Sun Solaris servers loaded with Oracle 7.x or on Windows NT servers loaded with Microsoft SQL Server. Private directories of IP addresses can be created and mapped to employee and customer names. NetSpeak also offers a Net-to-PSTN gateway that allows a company's employees to place calls to the public switched telephone network.

The Global Exchange Carrier Company (GXC) has also developed a gateway. Unlike the products of VocalTec and NetSpeak, which are used by companies to provide Internet or intranet calls for their employees, GXC is a public gateway and service that allows anyone on the Internet to make telephone calls. All you need in the way of software is VocalTec's Internet Phone and the GXC's Global Exchange (Gx) Phone. GXC uses the VocalTec Internet Phone as a voice engine, but implements the gateway function with its own software.

Using the company's Gx-Phone, which can be downloaded from its Web site, you go into the Internet mode, dial your destination number on the keypad, and click on the Send button. You'll be connected to the GXC gateway which completes the call to the destination party on an ordinary telephone line.

Among the features offered by GXC is on-demand calling records. All calls are logged to your account and can be retrieved for your records. GXC also offers directory assistance. Upon entering a person's name and location, you'll receive the most likely matches from GXC's Directory Assistance database.

GXC offers five calling plans as follows:

Calling plan	Name	U.S.A. (non-800)	U.S.A. (800)	Directory assistance	Account registration	Monthly subscription
1	GXC Basic	$0.25	$0.15	$0.25	$20	$0
2	GXC Enhanced	$0.20	$0.13	$0.25	$20	$10
3	GXC Deluxe	$0.17	$0.10	$0.20	$20	$20
4	GXC World Dial	$0.15	$0.08	$0.20	$20	$40
5	GXC Reserved Line	($6.00 per 30 minutes, prescheduled)				

Source: GXC Web page at http://www.gxc.com/callplan.htm.

IDT offers an international calling service called Net2Phone. IDT has a successful history offering people innovative means to bypass traditional phone systems and make cheap international telephone calls. Starting in 1991, IDT's callback services have let customers in 80 countries dial into IDT's phone switch in the United States, hang up after one ring, and be automatically called back to receive a U.S. dial tone. Customers then complete the call into the United States or anywhere else around the world and are billed by IDT at a U.S. rate significantly lower than the rate charged by the overseas phone company where the call originates. By combining its callback and Internet access technologies, IDT can charge someone in Japan just $0.10 a minute to speak with someone in the United States.

To use Net2Phone, a customer needs an audio-equipped PC, IDT's proprietary front-end Net2Phone software, and an IDT or IDT-partner Internet account which will allow them to be billed at a rate of $0.10 a minute for completing a call into the United States.

Voice-Enabled Applications

Several vendors have embarked on a course that fully integrates their phoneware with other Internet applications. In its first release of its Internet phone product, Quarterdeck integrated WebTalk with its Quarterdeck Mosaic Web browser, giving users even more incentive to get on the Internet. In future releases, Quarterdeck intends to provide an *inking* capability that allows users to view and mark up the same document or spreadsheet and talk about it over the Internet before posting final copies on private Web servers or distributing it via e-mail. The object is not only to simplify the way people work, but to help them complete transactions in real time. Not only can the participants benefit from a single graphical user interface to carry out multiple tasks, but there will be no additional overhead expense in doing so.

NetManage is thinking along the same lines as Quarterdeck but is approaching things in a different way. The company licensed its InPerson desktop videoconferencing application from Silicon Graphics in 1994, which is now bundled with other Internet applications within its Chameleon Enterprise for Windows. InPerson includes a whiteboarding capability that allows Silicon Graphics and Windows users to collaborate on projects using real-time desktop conference sessions. In addition, the package supports HyperText Markup Language 3.0 extensions in Web-Surfer, the company's Web browser.

Netscape Communications has embedded a real-time audio and video framework into its product line. Netscape Navigator 3.0 includes new telephony and whiteboard capabilities. In addition to voice, the telephony capability (Fig. 2.11) lets users send and receive text messages, a Web-based phone book helps users locate one another for conferencing, and an answering machine records messages and caller information. The whiteboard (Fig. 2.12) enables users to view and edit graphic images in real time using sophisticated drawing, zoom, and markup tools.

Microsoft has also embraced the concept of collaboration over the Internet; its NetMeeting is a true collaborative tool that combines the use of audio and data. It requires Windows 95 and has the following key features:

■ *Advanced calling features.* Allows you to specify one of three types of calls: audio only, audio and data, or data only (Fig. 2.13).

■ *Real-time voice.* Enables you to talk to another person over the Internet (Fig. 2.14) and can be used in conjunction with the other data and conferencing features.

■ *Application sharing.* Enables you to share a program running on one computer with other people in a conference. Up to 255 people can join a conference, pass around the virtual mouse, and share a Windows application.

■ *Whiteboard.* Enables you to sketch an organization chart, draw a diagram, type action items, and perform other tasks. You can point out your coworkers' errors by using a remote pointer or highlighting tool or take a snapshot of a window and then paste the graphic on a page.

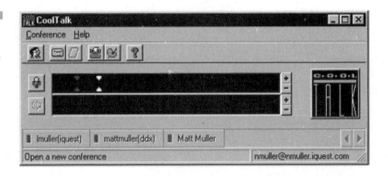

Figure 2.11
Netscape's CoolTalk interface, supplied by InSoft, allows you to make phone calls over the Internet by typing in another person's e-mail address or by selecting someone's name from a directory.

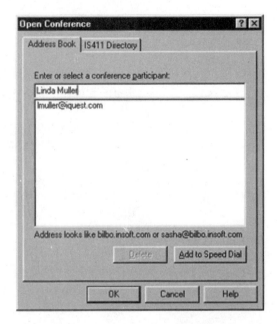

- *File transfer.* Enables you to send a file to all the participants in a conference by dragging it into the Microsoft NetMeeting window. The file transfer occurs in the background as everyone continues sharing programs, using the whiteboard, and chatting.

- *Chat.* Enables you to record meeting minutes and action items or carry on a conversation with fellow conference participants.

- *Online directory.* Enables you to easily find other people to communicate with on the Internet.

In early 1996, AT&T and IBM announced plans to get into Internet telephony. At this writing, IBM is farthest along in product development, having intro-

Figure 2.12

Netscape's white-board interface, part of InSoft's CoolTalk, allows you to engage in collaborative data sharing. In addition to all kinds of drawing tools, you can copy data from other applications and paste it into the whiteboard. In this case, a portion of Netscape's help file has been captured and pasted into the whiteboard. Then a comment was typed underneath, hinting to Ed that he should read the help file next time he doesn't understand something.

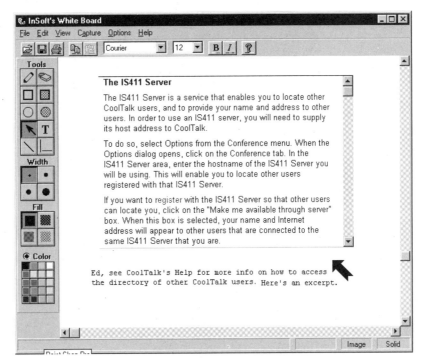

duced a beta version of its Internet Connection Phone in mid-1996 (Fig. 2.15).

A nice feature of the Internet Connection Phone is the search capability that is available when you call up the list of active users (Fig. 2.16). You can search for specific people or domain names in e-mail addresses.

A commercial version of the software will come preloaded on IBM multimedia PCs starting late second quarter 1996. It will eventually be included in IBM's OS/2 Warp operating system, as well as other IBM software and hardware products. Among the noteworthy technologies embedded in the Internet Connection Phone is an IBM-modified GSM compression algorithm that suppresses echos and better controls the loss of packets. According to IBM, the new algorithm compresses 8-kHz 16-bit voice samples to 9400 bits per second (bps) leading to clear, near echoless conversations.

One version of Internet Connection Phone takes full advantage of IBM's MWave—a technology that more efficiently processes multimedia and audio data whenever it can. A computer that has an MWave sound card installed can offload the Internet Connection Phone's compute-intensive compression and decompression, allowing the computer to do

Chapter Two

Figure 2.13
NetMeeting allows
you to specify the
type of call: audio
only, audio and data,
or data only.

other tasks more effectively while Internet Connection Phone is working.
Internet Connection Phone also comes in a version without MWave
sound card support.

The involvement of AT&T and IBM in Internet telephony not only
legitimizes this kind of activity, it will accelerate application development,
particularly for electronic commerce. Instead of merely posting a product
catalog on the Web, potential purchasers will soon be able to talk with a
live operator to get more information. After the purchase, users can access
the company's help desk page on the Web and speak with a technician for
configuration assistance.

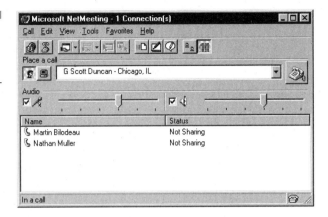

Figure 2.14
NetMeeting's real-time voice (telephone) interface.

Online Resources

The following table provides the Web links of the major phoneware vendors from which you can download working or evaluation copies of their software. The Web pages also contain such information as platforms supported, system requirements, product features, installation instructions, and troubleshooting advice. Since the technology is moving rapidly, you may want to consult these Web pages periodically for the latest developments.

Figure 2.15
IBM's Internet Connection Phone (beta release).

Developer	Product	Web page or FTP site
Camelot Corp.	DigiPhone	http://www.digiphone.com
Michael Cummings	NET-FONE (a handset)	http://www.net-fone.com
Engineering Consulting	ClearPhone	http://www.kaiwan.com/~radiobob/
FreeTel Communications	FreeTel	http://www.freetel.com
Global Exchange Carrier Company	Gx-Phone	http://www.gxc.com
IBM Corp.	Internet Connection Phone	http://www.internet.ibm.com/icphone/index.html
INRIA	Free Phone	ftp://zenon.inria.fr/rodeo/fphone
Intel Corp.	Intel Internet Phone	http://www.intel.com/iaweb/cpc
IRIS Systems	IRIS Phone	http://www.front.net/irisphone/iris.htm
JABRA Corp.	JABRA Ear PHONE (an earpiece)	http://www.pulver.com/jabra/index.html
Charley Kline and Eric Scouten	Maven	ftp://ftp.inet/pub/software/mac/voice/maven-20d37.hqx
Microsoft Corp.	NetMeeting	http://www.microsoft.com/ie/conf/
Netscape Communications Corp.	CoolTalk	http://www.netscape.com
NetSpeak Corp.	WebPhone	http://www.netspeak.com
MIT	PGPfone	http://web.mit.edu/network/pgpfone
Quarterdeck	WebTalk	http://www.quarterdeck.com
Hani Abu Rahmeh	Internet Multimedia	http://www.simtel.net/pub/simtelnet/win95/inet/immv110.zip
SilverSoft	SoftFone	http://www.pak.net/softfone.htm
John Walker	Speak Freely	http://www.fourmilab.ch/speakfree/unix/sfunix.html
Telescape Communications	TS Intercom	http://www.telescape.com/html/main.htm
Tribal Voice	PowWow	http://www.tribal.com/powwow/
VDOnet Corp.	VDOPhone	http://www.vdolive.com/vdophone
Voxware	TeleVox	http://www.voxware.com
VocalTec	Internet Phone	http://www.vocaltec.com
White Pine Software	CU-SeeMe	http://www.wpine.com
Van Jacobson and Steve McCanne	vat	ftp://ftp.ee.lbl.gov/conferencing/vat/alpha-test/vatbin-4.0b2-win95.zip

NOTE: This information, as well as updates, can be found at http://www.ddx.com/mgh.shtml.

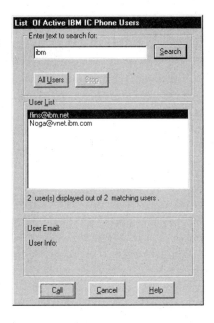

Figure 2.16
In this search within the Internet Connection Phone's list of active users, two users with the ibm domain name are online.

Conclusion

The key issue that must eventually be addressed by the vendor community is that of interoperability between the various phoneware products, especially if voice communication and voice-enabled applications are to migrate into mainstream business use. After all, the conventional telephone's success has depended in large part on the interoperability of different vendors' phones, switches, and software all working together and in conformance to a uniform set of standards.

The Internet Telephony Interoperability Project at MIT has been established to work with research partners toward achieving interoperability among Internet telephony applications. Among its research activities is the exploration of opportunities for increasing interoperability between the Internet and the public switched telephone network. More information about the Internet Telephony Interoperability Project can be found at

http://rpcp.mit.edu/~itel/

Intel has taken an important step in the direction of interoperability with its Intel Internet Phone for Windows 95 (Fig. 2.17), which interoperates with H.323-based communications software from Microsoft and

Figure 2.17
Designed as an add-on helper applet compatible with Microsoft Internet Explorer and Net-scape Navigator browsers, Intel Internet Phone can be configured to snap into the bottom edge of the user's Web browser.

other vendors that have committed to supporting this standard. H.323 is a worldwide standard that specifies how audio and video should run over packet-switched networks, such as the Internet. Future versions of Microsoft NetMeeting, for example, will support H.323. This means users of Intel Internet Phone and Microsoft NetMeeting will be able to talk with each other.

Intel Internet Phone also incorporates Microsoft's User Location Service (ULS) technology to locate other Internet phone users through existing Internet Directory Services. ULS is supported by leading Internet directory services—such as Four11, Banyan's Switchboard, WhoWhere, DoubleClick, and BigFoot—making it easy to find and call other users of H.323-based Internet Phones. Once connected via an Internet phone on a multimedia PC, users can talk to each other over the Internet while using other Internet applications.

The technology that enables phone calls to be made over the Internet is changing rapidly. Not only are new features continually being added, but major advancements are being made in the area of voice compression. The quality of voice over the Internet will be significantly improved with the addition of transport protocols that handle real-time traffic. In the very near future, Internet telephony and associated applications for collaboration will make the Internet as crucial for daily business communications and personal empowerment as the telephone, but far more compelling and engaging.

CHAPTER **3**

Videoconferencing Over the Internet

Introduction

In recent years, videoconferencing has become an increasingly accepted form of communication among businesses and government agencies that want to save on travel costs, encourage collaborative efforts among staff at far-flung locations, and enhance overall productivity among employees. Originally, videoconferencing was seen as a method to link people at remote locations over wide area networks (WANs). More recently, videoconferencing is being used to link desktop computer users over local area networks (LANs) in an effort to obtain the same benefits within a building or campus environment.

Today, the technology has progressed to the point where you do not need access to expensive high-speed networks to enjoy the advantages of videoconferencing with friends, relatives, and coworkers. Now, with some relatively inexpensive hardware and software, virtually anyone can participate in videoconferences over the Internet. You can conference as long as you want with anyone, anywhere in the world—at little or no cost—provided that the other participants have a comparably equipped computer—sound board, speakers, microphone, video camera—the same videoconference software, and an Internet connection.

A videoconference works much like the phoneware discussed in the previous chapter, except that a video component is now added to the transmission. To start a videoconference session over the Internet, both parties need to be online at the same time with their multimedia-equipped computers. You open the videoconference program and simply type in the address of the other party, and the software makes the connection.

Once a connection is made, each person can see the other and speak into the microphone to record and transmit his or her voice as digital data through the Internet. There are two distinct coding/decoding processes involved in supporting a videoconference: one for voice and another for video. When the data is received at the other end, the voice stream is immediately decompressed and reproduced through the destination computer's sound board and speakers. In addition, the video is received, decompressed, and displayed in a remote window. All this happens very quickly, starting as soon as either party starts speaking into the microphone.

The key to a natural-sounding conversation is to minimize the delay involved in the process of compression/decompression. The codec algorithms used by various vendors differ in the level and speed of compression they offer, which determines voice quality. The better the sound quality, the more processing power it takes to make the transmission smooth and natural. If a high-quality codec is used with a low-performance computer, sound will suffer because the machine cannot keep up with the volume of data it is trying to produce.

An example of a high-quality codec is the one used by Cinecom Corporation's CineVideo/Direct. The company uses a modified version of GSM (Global System for Mobile communications), an international standard for digital cellular communications which is especially popular in Europe. Since Cinecom has optimized GSM for use on the Internet, it can only be used in conjunction with CineVideo/Direct. Not only does this make CineVideo/Direct incompatible with the videoconference software of other vendors, it is not even compatible with other GSM standard devices such as digital cellular phones.

Admittedly, image quality over the Internet will not be the same as the image quality that can be achieved over high-speed LANs and digital WAN services such as ISDN (integrated services digital network)—at least for now. No matter what the vendors' advertising may say, video over the Internet tends to be grainy and the picture itself is very small compared to the available screen space on your monitor. The resolution of some videoconferencing products is limited to 160 by 120. Black-and-white displays are generally sharper than color images, but still usable. If you try to adjust the image window to make the picture larger, the graininess will worsen until the picture becomes so pixelized that the image is no longer comprehensible. You should also be aware that the audio component of the conference tends to run ahead of the video component; that is, the speakers' lips and body movements will not be synchronized with the person's voice. So what you will get is a videoconference that looks and sounds like a badly dubbed Godzilla movie.

There are several reasons for all this. The Internet is based on a bunch of protocols collectively known as TCP/IP which is short for *Transmission Control Protocol/Internet Protocol*. These protocols were designed for text messaging and file transfers—not the real-time multimedia applications that are becoming so popular today. TCP was designed to deliver conventional data reliably, but in bursts, with delivery delays that are not acceptable for real-time delivery of audio and video, where data must be received in a continuous stream. Delay has several causes.

One source of delay comes from the processing of audio and video signals by your computer. There is a certain amount of delay while your computer samples and compresses the audio and video signals and packetizes them for transmission over the Internet. The audio component is sampled and put into digital form and compressed, as is the video component, via coder/decoders (codecs). Two different compression algorithms are used: one for voice and one for video, with video compression consuming the lion's share of your computer's processing resources. Most of the videoconferencing products for the Internet use software-based codecs, which means that processing delay will be significant. Decompressing the voice and video at the receiving end is less of a problem and goes much faster.

Another source of delay is the Internet. Not only do the packets take time to travel over the Internet—going from server to server until they reach their destination—they may take different paths to the destination and may not all arrive in time to be reassembled in the proper sequence. If this was ordinary data for an e-mail application, for example, late or bad packets would simply be dropped and error-checking protocols would request a retransmission of those packets. But this concept cannot be applied to packets containing compressed audio and video without causing major disruption to the conference session, which is supposed to be occurring in real time. As a result, late packets are simply dropped, which causes gaps in speech and choppy video.

A key disadvantage of the Internet Protocol (IP) is that it does not allocate a specific path or amount of bandwidth to a particular session. The resulting delay can vary wildly and unpredictably, wreaking havoc with real-time applications such as videoconferencing. The distance between videoconference participants also contributes to delay, as does the traffic load on the Internet at any given time of the day.

A number of solutions have been proposed to address the problems of TCP/IP for multimedia applications and some of them are already being implemented on the Internet. Two of them are the resource ReSerVation Protocol (RSVP) and the Real Time Protocol (RTP). Together, these and other protocols will soon greatly improve the performance of videoconferencing and other multimedia applications running over the Internet.

To remedy the shortcomings of IP, what is needed are resource setup functions similar to those used in the public telephone network, where connections stay in place for the duration of a conversation. One of the most promising of these is RSVP, the resource ReSerVation Protocol, developed by the Internet Engineering Task Force (IETF). RSVP runs on top of IP to provide receiver-initiated setup of network resources on behalf of an application data stream. When an application requests a specific quality of service for its data stream, RSVP is used to deliver the request to each router along the path of the data stream and to maintain router and host states to support the requested level of service. In this way, RSVP essentially allows a router-based network to mimic the dedicated connections of a circuit-switched network on a best-efforts basis.

RTP was developed by the IETF as an alternative to TCP. RTP works alongside TCP to transport streaming data across networks and synchronize multiple streams (e.g., so that audio matches the video). RTP inserts timing and sequencing information into each packet to minimize delay and variable delay. Although RTP does not enhance the reliability of a transmission the way RSVP does (i.e., by reserving enough bandwidth to support the application), applications make use of the timing and sequencing information provided by RTP to enable audio and video streams to run smoothly, despite occasional dropped packets.

Still another source of poor-quality video is the low frame rate that is typically used in videoconferences over the Internet—usually less than 2 frames per second, versus 30 frames per second for broadcast quality video. The lower frame rate is necessary to accommodate the 28.8 Kbps and lower speeds at which most people connect to the Internet with their modem-equipped computers. Of course, higher-quality videoconferencing is possible over TCP/IP-based private intranets which use dedicated high-speed links between corporate locations. The use of all-digital end-to-end services such as ISDN also allows for higher-quality videoconferencing, but at a much higher connection cost.

Although you should be aware that these alternatives exist, further discussion of these topics is beyond the scope of this book, which is oriented to achieving free or low-cost communication over the Internet without the need for expensive hardware or high-speed networks. Of note is that some software, such as VideoPhone from Connectix Corporation, supports videoconferencing over virtually any kind of network, including those based on Ethernet, token ring, ISDN, FDDI (Fiber Distributed Data Interface), frame relay, and ATM (Asynchronous Transfer Mode). And despite the current limitations of running a videoconference over the Internet,* the software is cheap and easy to use. It works fine for casual communication with family and friends—and you can talk for as long as you want with people anywhere in the world at little or no extra cost for your basic Internet connection. You can even record video messages offline and send them as MIME attachments via e-mail (Fig. 3.1). Upon playback by the recipient, the video will exhibit much better quality than if it were sent live over the Internet. This topic is covered in more detail in Chap. 5 on "Voice Mail and Video Mail Over the Internet."

There are three kinds of videoconferencing: (1) *point to point,* in which you communicate with one other person via the Internet or a private net-

---

* Connectix admits that its VideoPhone software will work over the Internet, but slowly. The company says it is not directly marketing VideoPhone to this market segment because it feels customers may have unrealistic expectations. VideoPhone works best over ISDN or faster links. Nevertheless, VideoPhone is very inexpensive and works well enough over the Internet for casual use.

Figure 3.1
With Connectix
VideoPhone, for
example, you can
record a video mes-
sage and send it as
a MIME attachment
via e-mail.

work; (2) *group,* in which you join an existing conference by connecting to a special server equipped with reflector software; and (3) *one-way,* in which you receive broadcast data from a server. The *reflector* is server-based soft-ware—available in UNIX, Windows NT, and Windows 95 versions—that allows users to conduct and manage meetings over the Internet, private TCP/IP networks, and the MBone. It enables multiple users to conduct multiple video/audio and chat group discussions, collaborate over docu-ments with a whiteboard, and *cybercast* (broadcast) to large audiences. Among the companies that offer this kind of software is White Pine Soft-ware, whose reflector performs such tasks as network bandwidth control, video pruning, audio prioritization, and conference management.

With its automatic bandwidth detection capability, for example, White Pine's reflector accommodates the various bandwidth needs of individual users, while balancing the needs of the network system as a whole. For example, the reflector can adjust transmission rates of specific users on-the-fly if packet loss is running too high due to heavy network traffic. If network traffic is too heavy for a reliable conference, the transmission rate will remain at the lowest setting and only move up when the network is less congested.

Conference management is also done at the reflector. From the reflec-tor, a moderator can manage viewers—whether they are visible, hidden, or lurkers (explained later). The moderator can control access to the confer-ence, single out individuals for a conversation, turn group microphone and speakers on and off, and identify the current speaker.

Internet users establish a connection to the reflector to videoconfer-ence with others. However, the reflector software is available for purchase for installation on private servers to enable organizations, educational institutions, and end users to set up their own multiparty conferences and live broadcast stations.

The last kind of videoconference—broadcast—is not really conferencing at all because it does not involve two-way interaction. Instead, it allows one person to deliver a multimedia presentation to one or more people simultaneously. This topic will be discussed separately in Chap. 8, "Broadcasting Over the Internet." However, there are products that support all three modes of video, such as Enhanced CU-SeeMe* from White Pine Software, while others such as Connectix's VideoPhone only support point-to-point calls and broadcasts. Both products provide a whiteboard for data sharing that allows users to see documents and take turns marking them up for revision.

System Requirements

The hardware requirements for videoconferencing are the same as those for placing telephone calls over the Internet, with the addition of a video camera. Instead of phoneware, you will need special videoconferencing or video phone software—the two terms are often used interchangeably, but the term video phone implies a two-party connection—that provides the algorithms that reduce the amount of data to the bare essentials so face-to-face meetings can be conducted over relatively low bandwidth Internet access connections.

To run the software-based codec on a PC, a 486 or better processor is required; when used on a Macintosh, a 68020 processor or better is required. For the best quality video, you will need a modem that operates at 28.8 Kbps or better. For audio-only conferencing, a 14.4 Kbps will do. Regardless of computing platform, you will need between 5 and 10 MB of available disk space for a full installation of most videoconference programs. You will also need at least 8 MB of RAM but, of course, more is better, especially if you're using Windows 95. As with placing telephone

---

* Enhanced CU-SeeMe is the commercial version of the original CU-SeeMe freeware developed at Cornell University (CU). The commercial version, which is fully compatible with the freeware version, includes additional features to make videoconferencing easier to use and manage, such as a higher-quality, low-bandwidth encoding method; a whiteboard capability for collaboration during conferences; Web browser support for launching Enhanced CU-SeeMe from a Web page; a phone book to manage conference addresses and for direct-dialing conferences; and a text-based chat facility.

calls over the Internet, videoconferencing over the Internet requires a sound board, and microphone with speakers or headphones. However, an additional piece of equipment you will need for videoconferencing is a video camera.

There are video cameras for gray-scale or color image input. Gray-scale cameras generally offer a sharper image and the use of black-and-white images requires much less bandwidth than color. You can spend as much as you want on a video camera for your computer—as much as $1000 and more. Among the least-expensive video cameras are the Connectix Quick-Cam (less than $100 for gray scale and less than $200 for color), the Video-Labs FlexCam (less than $200), and the Toshiba IK-M28 Desktop Video Camera (less than $300).

Some videoconferencing products require a video-capture board that the camera connects to so you can send and receive audio and color or gray-scale video. Such boards compress the video signals, relieving the computer's CPU of this burden. Among the most popular video-capture boards are Creative Labs' RT-300, Intel's Smart Video Recorder, and Logitech's MovieMan. These boards cost less than $300 through mail order, but other higher-priced boards are available that offer a complete video capture, edit, and compression solution based on industry standard MPEG. These boards can sell for $900 and more.

However, Connectix's QuickCam does not require a video-capture board. The sub-$100 camera simply plugs into your computer's parallel port and video compression is done in software. (You could also plug the QuickCam into the video input port of a video board.) While simple and inexpensive, this solution does have two key disadvantages: (1) if your computer has only one parallel port, you can't use your printer and the video camera at the same time; and (2) since video compression is done in software, you must have at least a 486DX2 computer to handle the processing load.

You could also get a good deal on hardware and software by buying a bundled package that provides a complete desktop videoconferencing solution. For example, White Pine offers its Enhanced CU-SeeMe for Windows and a VideoLabs FlexCam full-motion color video camera, which includes a Stinger video capture and overlay board. A less-expensive bundle offers the Connectix QuickCam with Enhanced CU-SeeMe. The advantage of buying bundled videoconferencing solutions is that the vendors have done all the hardware-software compatibility testing and fine-tuning for you, eliminating many of the potential problems that often arise when you try to mix and match products yourself.

Instead of buying a new video camera for your computer, you could use your camcorder if it has an audio/video (AV) out connector. You would attach one end of the audio/video output cable to the camera's AV out connector and the other end to the video board's video-in and audio-in ports (Fig. 3.2). You can even mount the camcorder on a tripod and use a remote control for special effects such as zooming in and out and fading in and out.

You may have heard of a new type of modem known as DSVD (Digital Simultaneous Voice and Data). This is a special type of modem that supports the simultaneous transmission of voice and data over a dial-up line. DSVD modems have external audio input and output connectors and handle all the voice compression and decompression, as well as transmission. Since the PC never has the opportunity to manipulate the voice data, the voice function of the modem is totally independent of any application on the computer. This has several advantages: with its own codec, it relieves some of the processing burden on the computer's CPU; with its own audio input and output connectors, it eliminates the need for a separate sound board. In addition, there is no longer any need to interrupt telephone conversations to transmit data, which eliminates the expense of a second telephone line. Some DSVD modems, such as the Logicode QuickTel II-C 28X-SVD, even come complete with built-in microphone and speakers, eliminating the need to buy even more hardware. Other companies that make DSVD modems are Best Data, Creative Labs, MultiTech, US Robotics, and Zoom.

A newer version of the DSVD standard specifies a scheme for dynamically allocating bandwidth to data and/or voice as needed. For example, if you are busy browsing the Web at 28.8 Kbps and a voice call comes in, the data rate will step down to 19.2 Kbps so voice can be accommodated in the newly created 9.6 Kbps of bandwidth.

Although not required for telephone calls, faxing, and videoconferencing in cyberspace, a DSVD modem is a good choice if you have to upgrade your old low-speed modem, especially if you can get one that is upgradeable through firmware. In some cases, you can get free upgrades of features and speeds—28.8 to 33.6 Kbps, for example—by going to the vendor's Web page. These modems cost more, but you will not have to worry about product obsolescence for many years to come.

When installing the videoconference software, you will have to configure it to work with your system by specifying such things as modem speed, COM port, sound board variables, input device, and display type (Fig. 3.3) to ensure proper operation.

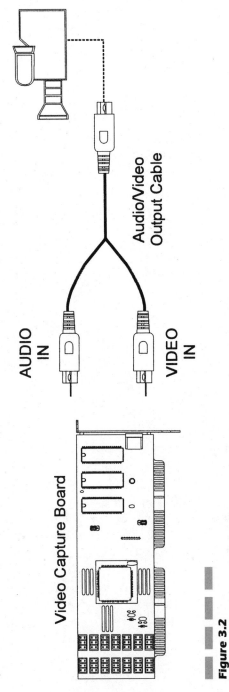

Audio/Video
Output Cable

AUDIO
IN

VIDEO
IN

Video Capture Board

Figure 3.2
Camcorder connections to video capture board. Refer to your camcorder and video board installation instructions for more detailed information.

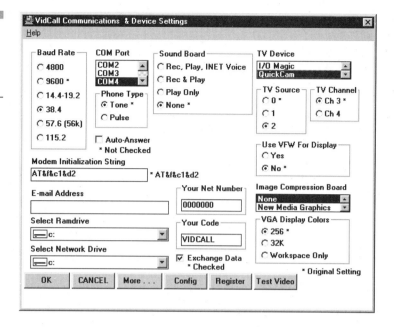

Figure 3.3

MRA Associates Vid-Call communications and device settings window.

Point-to-Point Conferencing

Point to point (person to person) is the simplest form of videoconferencing. In this type of conference, one person calls another person directly over the Internet with a dial-up modem connection. The two computers, both running the same type of software, are then connected.

The process is very simple. With Enhanced CU-SeeMe, for example, you make a person-to-person call as follows:

1. Choose Phone Book from the Conference menu or select the phone book icon, which is the first icon on the toolbar (Fig. 3.4).

2. Enter the IP address of the person you are calling. This address, assigned by the Internet service provider, has four sections separated by a dot. For example, the IP address of White Pine's Enhanced CU-SeeMe's demonstration conference is

    192.233.34.5

3. Click the Call button. The local video window status changes to "Connecting."

Figure 3.4
The Phone Book in
White Pine Software's
CU-SeeMe.

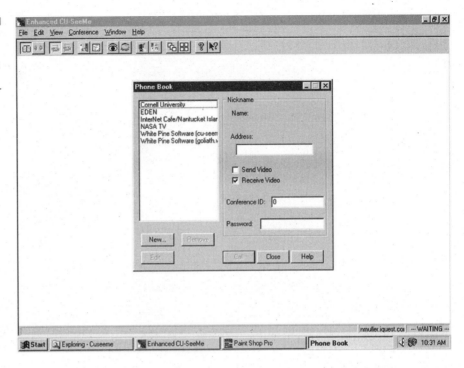

When you are connected to the person you are calling and video is being sent, his or her remote video window appears on your desktop. Most videoconference software allows you to view yourself off line so you can check to see how you look to other participants during a conference and do any necessary primping and preening before actually going online with the world (Fig. 3.5).

Group Conferencing

To conference with a group of people, whether over the Internet or a private intranet, all participants must meet at a reflector-equipped server. When using Enhanced CU-SeeMe over the Internet, you can try connecting to White Pine Software, where the company has a dedicated reflector set up for Enhanced CU-SeeMe users. There you will meet other Enhanced CU-SeeMe users, as well as some White Pine employees who can answer your questions about the product.

To conference with other Enhanced CU-SeeMe users at a reflector, the connection process is as follows:

Figure 3.5
Using Connectix
VideoPhone, for
example, you can
view yourself offline
to see how you will
look to other confer-
ence participants
before you go online.

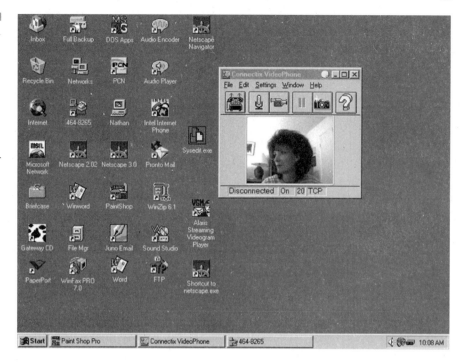

1. Choose Phone Book from the Conference menu or select the Phone Book icon on the toolbar.

2. Enter 192.233.34.5 or White Pine Software from the Phone Book.

3. Click the Call button. The local video window status changes to "Connecting." If a message appears indicating that there are too many participants, try again in a few minutes. After repeated unsuccessful connection attempts, you may have to click on the Disconnect button before trying to connect again.

4. After reading a Message of the Day from the reflector, click OK. You may also get a list of available conferences from which to choose.

Each person sending video to the White Pine Software reflector appears in a separate remote video window. You can have as many as eight remote video windows open on your desktop at the same time, but you can limit the number of windows to less by entering your preference in the Conferencing dialog box.

White Pine describes three types of conference participants: visible users, hidden users, and lurkers. A *visible user* is a participant who is sending a video image to the conference. Anyone in the conference can choose to see this video image. A *hidden user* is a participant sending a video image, but you are not currently seeing the image. To turn a hidden user into a visible user, click on that participant's entry in the Hidden User section of the Participant List. If you already have the maximum number of visible users displayed on your screen, you will need to turn a visible user into a hidden user by closing that participant's window or by clicking on that participant's entry in the Visible User section of the Participant List. A *lurker* is a participant not sending any video image at all. Perhaps the lurker's computer has no video capability, or the participant may have chosen to disable his or her video camera.

There are several types of controls with which you can make adjustments during the videoconference to show viewers or improve audio and video quality:

■ *Listener.* To alert you of incoming calls with caller ID, giving you the option of accepting or rejecting incoming calls.

■ *Video controls.* To easily find out the status of your connection, configure transmission and reception settings, and to set the brightness and contrast for your local video window.

■ *Audio controls.* To adjust the volume, turn on and off speaker and microphone, and to select either voice-activated communications or a push-to-talk option.

■ *Audio compression.* To select the most appropriate audio compression method of 2.4 or 8.5 KB to support 14.4 and 28.8 Kbps modems respectively; or 16 or 32 KB for higher bandwidth connections.

■ *Video compression.* To select standard or high resolution settings.

To talk to remote video users, you need to have a microphone. Click and hold the Push To Talk button in the Audio window. The button changes to Transmitting so you can talk to the other users. To talk directly to a particular user, click on the microphone icon in his or her window.

If you do not have a microphone or your connection is too slow to carry audio, you can still communicate with others using the text chat capability. This is activated by choosing Chat from the Conference menu or clicking the Chat icon on the main menu button bar. A chat window will open into which you type text. Pressing the Return key will send your message. If you have speakers or a headset, you will be able to hear other participants. You can type a message in the Chat window asking another participant to talk. Other participants who have their Chat windows open will see your message. Your text appears in the top half of the window with your name. Anyone who has a chat window open will see your text and you will see the text that other users are typing.

Demonstration versions of both Enhanced CU-SeeMe and White Pine Reflectors are available for electronic download from the company's Web site at

http://goliath.wpine.com/cudownload.htm

The 30-day demonstration versions will only run with a serial number provided to you via e-mail from White Pine Software.

Some publicly accessible reflectors through which you can participate in videoconference sessions with other CU-SeeMe users are as follows:

Organization	IP address
Cornell University	132.236.91.204
EDEN	199.171.21.1
Internet Cafe/Nantucket	204.249.164.2
NASA TV	139.88.27.43
University of Pennsylvania	130.91.72.36
Utopia	137.130.65.31
White Pine Software	192.233.34.5
White Pine Software	192.233.34.20

These sites are entered in the CU-SeeMe phone book. You just select the Conference menu and open Phone Book. A more comprehensive listing of reflector sites, grouped by Windows and Macintosh platforms, is provided by Michael Sattler, whose Web page is at

http://www.indstate.edu/msattler/sci-tech/comp/CU-SeeMe/reflectors/nicknames.html

If you have a Web page of your own, you can set up Enhanced CU-SeeMe to launch a conference via a hypertext link. The link goes to a file that defines the IP address and Conference ID of the reflector to which you want to connect. This file is placed in your server directory, along with your other HTML files. You must also configure your Web browser to launch CU-SeeMe automatically when the link is activated. Full instructions on how to do this are provided by White Pine Software at

http://support.cuseeme.com/whiteP/h168.HTM

Another videoconference product that makes use of a reflector component at a server is Cinecom's CineVideo/Direct. To simplify the process of connecting two parties together through the Internet, Cinecom maintains an online program called the CineVideo/Director. The Director sits on a dedicated server on the Internet or corporate intranet. Its main function is to connect people together, but it performs several management functions as well.

When you start up CineVideo/Direct, it sends a message to Cine-Video/Director with your configuration information—specifically, your e-mail address and current IP address. Anyone wishing to contact you simply types in your address and the Director handles the IP address connections. Once the connection is established, the Director drops out leaving the two parties to talk directly. The setup transaction is simple and brief, enabling the Director to handle a large volume of calls. The only thing you notice is confirmation that a connection has been made or a message replying that the user is either offline or busy.

Without the Director, each party would have to know the IP address of the other in order to connect directly. This is not a feasible solution because most people with dial-up access to the Internet are assigned an IP address at the time they log on. In fact, they may be assigned a different IP address each time they log on to the Internet. (See the discussion of dynamic versus static IP addresses in Chap. 2.) Even those people with fixed IP addresses probably do not know what they are. The role of CineVideo/Director is to provide online IP address matching for anyone using CineVideo/Direct, similar to the way some phoneware products require users to log on to a directory server before they can talk with another person directly. CineVideo/Director provides video yellow-page listings for up to 1000 simultaneous users and can be used to manage users logged on to private intranet servers as well.

Among the management features of CineVideo/Director is security. When a user accesses the Director, the user database is searched for the user record. The user record contains information about system access authorization. If the user has system access, then access is granted to the system. A lock and key mechanism is used to turn Director services on and off.

The Director allows the system administrator to create Video-on-Demand (VOD) files/folders on the server, which authorized users can access. These VOD files can consist of announcements, product demonstrations, training modules—virtually anything a person or a company has recorded and stored for on-demand viewing. The VOD files themselves are created by compressing AVI files into CineVideo format, which provides compression ratios as high as 10:1, compared with an AVI file. File compression varies, depending on the audio and video content.

The system administrator can provide a description of the VOD file which is shown on the Director's video services menu and assigns a security key to control access to the VOD files. VOD folders can also have security keys to prevent unauthorized users from even knowing that certain folders or files are available for viewing.

The Director also has a broadcast capability that provides one-to-many distribution of video, audio, and text. The originator of the material broadcasts data into the server where the Director replicates it for distribution to other users who are also connected to the server to receive the broadcast. With this type of connection, video and audio are sent in one direction—from the broadcaster to the receivers. However, text can be transmitted in both directions, allowing receivers to interact with each other and the broadcaster via the chat capability.

The Director's ODBC (i.e., Microsoft's Open Database Connectivity specification) database is used to keep all system accounting information (Fig. 3.6). This includes user information and user system configurations. Information about number of accesses, total time used, last time and date on, e-mail address, and organization/company are stored in the user record. Any ODBC-compliant product such as Microsoft Access or SQL Server can be used to manipulate this database for report generation or other database functions.

FreeVue Telecommunications Network offers a slightly different way for users of its FreeVue videoconference software to find each other. In addition to providing free software with which you can videoconference with others, the FreeVue software allows you to make Internet phone calls, communicate in text mode with the chat facility, watch live Internet broadcasts, and create your own television station.

Figure 3.6
Cinecom's CineVideo/ Director provides an ODBC-compliant database to keep all system accounting information. This screen shows a usage report provided by the Director.

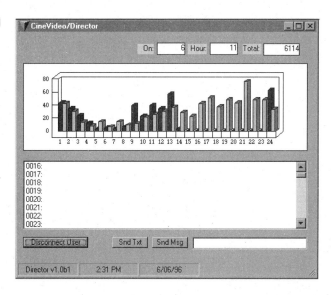

The company provides directory information on its Web page. In addition to a complete listing of other FreeVue users, you can find others to conference with by initiating a search of the White Pages or Yellow Pages. With the White Pages, you can search by online name (handle), real name, city/province, state, zip code, country, or e-mail address. With the commercially oriented Yellow Pages (Fig. 3.7), you can search by company name, company category (arts/entertainment, business/finance, sports, computer related, Internet broadcasters, and other), online name (handle), and e-mail address.

FreeVue also offers a Web page that lists active users. With FreeVue running, you can click on a name to be directly connected. You can also conference with multiple people by clicking on multiple names. Pressing the update button at the end of the list refreshes the screen with the latest information as people log on and off.

MRA Associates allows users of its VidCall to set up their own video-conference group over the Internet using the company's WhoIsThere software, which you download from the company's Web page and unzip into the same directory as VidCall. The conference group becomes visible when clicking on QUICK CALL|WhoIsThere while in VidCall. If the Vid-

Figure 3.7

FreeVue Telecommunications Network provides directory information on its Web page, including a search capability for its White Pages and Yellow Pages.

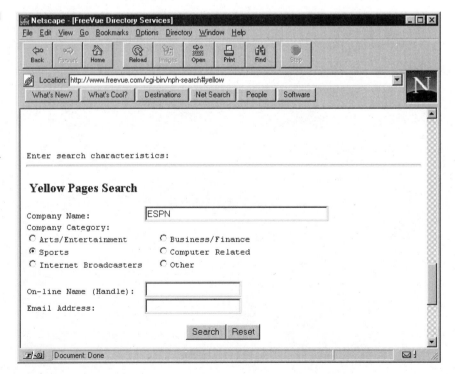

Caller is logged on to the Internet, logging on to a specific conference group identifies the caller to all others who are logged on and available to participate in a VidCall session.

Features

Videoconference products differ in terms of performance and features. After narrowing down your choices to a few products based on how well the vendors' evaluation software performed, you can turn your attention to comparing features to make a final selection. Here is a list of features you may want to look for:

- *Address books.* A facility that allows you to keep addresses and other information about people you conference with frequently. Clicking on the address is usually all that is needed to establish the connection, provided you are already logged on to the Internet and registered with a directory server, if required.

- *Audio compression control.* The ability to select the audio compression method, depending on the speed of the computer. For example, a 486-class machine might have to use ADPCM for audio compression, whereas a Pentium-class machine can use GSM which is more processing-intensive but offers better-quality audio.

- *Audio messaging.* Allows you to record and send an audio message to one or more people (also called *voice mail*).

- *Auto answer.* Automatically answers incoming calls.

- *Broadcast.* A capability that allows you send video (or audio or text) to multiple receivers simultaneously. Usually the actual broadcast is handled by a specially equipped server where other users connect to receive the replicated data stream. These data streams are delivered to multiple locations over diverse paths.

- *Browser configuration.* You can set your browser to use the videoconference program as a viewer when you encounter links on the Web that use the file extension supported by the conference program. For example, Connectix VideoPhone uses the CVP file extension. When you click on a link with an embedded CVP file extension, the browser automatically opens VideoPhone to view the file.

- *Bye button.* Allows you to hang up on a single caller using a single button.

- *Call hold.* Allows you to put the current videoconference call on hold so you can answer another incoming call.

- *Call limit.* You can set a limit on the number of video calls you can take at once.

- *Call notification.* Provides a ring, tone, or announcement of an incoming call, allowing you to decide whether to accept or reject it.

- *Chat.* A facility that allows someone to participate in a conference without necessarily having a sound board and microphone or video camera. All participants can see and respond to the typed messages through a separate chat window that shows up on their screens.

- *Clipboard.* A facility for capture and temporary storage of any application of any window.

- *Cross-platform communication.* The capability to videoconference across computer platforms (i.e., between users of PC and Macintosh and UNIX machines).

- *Document markup.* Allows conference participants to make notations on documents viewed through the whiteboard facility.

- *File transfer.* Allows participants to send files to each other during the videoconference.

- *Frame rate control.* The ability to specify the frame rate (in frames per second) at which the video camera operates, keeping in mind that higher frame rates will impose a greater processing burden on the computer and will require more bandwidth during transmission. If either or both types of resources are not enough, conference quality will be greatly diminished. While a frame rate of 7 to 10 frames per second results in acceptable quality, realistically you can expect only 2 frames per second due to the delay characteristics of the Internet.

- *Graphics drawing.* Allows conferees to draw lines, arrows, rectangles, and circles in the whiteboard.

- *Group conferencing.* Allows you to conference with multiple people at the same time. Each person is viewed in a separate window on your desktop. Usually, the limit is eight people.

- *Image support.* A capability that allows you to grab a video image and associate it as a still shot with an entry in the address book.

- *Import facility.* Imports text and graphics into a whiteboard.

- *ISDN connection.* A capability that permits a videoconference session to be set up over all-digital ISDN lines for the best quality video and audio. This type of connection is ordered from the local telephone company and requires a terminal adapter instead of a modem.

■ *LAN connection.* A capability that permits a videoconference session to be set up between participants connected via a local area network such as Ethernet or Token Ring.

■ *Message playback.* The ability to review recorded audio or video messages before deciding to send them out.

■ *Modem connection.* A capability that permits a videoconference session to be set up directly between two participants over ordinary phone lines instead of over the Internet.

■ *Notation transmittal.* The ability to transmit the notations of participants that have been collected in the whiteboard so that everyone can have a copy of what was done during the conference.

■ *Online help.* Entails access to reference materials and assistance from within the videoconference application. Windows-based online help facilities usually include a contents page and search facility.

■ *Picture control.* Allows you to specify image quality by adjusting contrast, tint, saturation, and brightness.

■ *Private log in.* Allows you to log in to a server without your name appearing in the directory listing of active users so you can avoid unwanted callers.

■ *Resolution control.* Allows you to set the resolution for the videoconference window. The lower the resolution (e.g., 120×90 pixels), the clearer the picture; the higher the resolution (e.g., 320×240 pixels), the grainier the picture.

■ *Search.* Allows you to search callers listed in your address book (or online at a directory server), sometimes by entering only the first or second letter.

■ *Snapshot.* Permits a conferee to capture a region of the window and place it in the whiteboard for viewing by all participants.

■ *Video compression control.* The ability to select the video compression method, depending on the speed of the computer. For example, with a 486-class machine running Windows 95 and connected to the Internet, you might set the compressor for full frames (uncompressed), but not specify compression quality or *interframe compression* (increases the motion seen over low-bandwidth links by sending only the differences in consecutive frames). With a Pentium-class machine, you can compress the frames, specify compression quality (level of graininess) and interframe compression because these tasks are more processing-intensive.

■ *Video mail.* Allows you to record and send video messages to one or more people.

- *Video for Windows API.* A Microsoft application programming interface (API) that enables the videoconference program to accept any Video for Windows—compatible capture device as the input source.

- *Video-on-demand (VOD).* The capability to send video to a server where it can be accessed on demand by authorized users.

- *Whiteboard.* A facility that allows you to share information with other conference participants. This facility usually includes drawing and editing tools, and allows you to save the contents of the workspace to a file which can be viewed again in the whiteboard.

Online Resources

The following table provides the Web links of the major Internet videoconference software vendors from which you can download working or evaluation copies of their software. The Web pages also contain such information as platforms supported, system requirements, product features, installation instructions, and troubleshooting advice. Since the tech-

~Developer	Product	Web page or FTP site
Cinecom Corp.	CineVideo/Direct	http://www.cinecom.com
Connectix Corp.	VideoPhone	http://www.connectix.com
FCS	VideoVu	http://www.totw.com/videovu.htm
FreeVue Telecommunications Network	FreeVue	http://www.freevue.com
INRIA (Institute National de Recherche en Informatique et en Automatique)	INRIA Videoconferencing System	http://www.inria.fr/rodeo/ivs.html
MRA Associates	VidCall	http://www.access.digex.net/~vidcall/vidcall.html
NetManage	InPerson (included in the Chameleon Desktop product line)	http://www.netmanage.com
SmithMicro Software	AudioVision	http://www.smithmicro.com
VDOnet Corp.	VDOphone	http://www.vdolive.com/vdophone
White Pine Software	Enhanced CU-SeeMe	http://www.wpine.com

NOTE: This information, as well as updates, can be found at http://www.ddx.com/mgh.shtml.

nology is moving rapidly, you may want to consult these Web pages periodically for the latest developments.

Conclusion

Just as more Web applications are now becoming voice-enabled, we can expect the next generation of Web applications to become video-enabled. This will spur new innovations in such areas as customer service, technical assistance, training, corporate recruiting, news services, and electronic commerce. As the Internet continues to grow in terms of subscribership and global reach, businesses will have every incentive to differentiate themselves from competitors by adopting video technologies that help to convey an up-to-date, vibrant corporate image.

As more bandwidth is added to the Internet and the use of new protocols becomes more widespread, videoconferencing becomes increasingly more feasible. Continual improvements in software-based codecs, modems, and sound boards will enhance the videoconferencing experience through better audio and video quality. Faster Internet connections available through ISDN, cable, and emerging local loop technologies such as ADSL (Asymmetric Digital Subscriber Line),* will soon bring low-cost, CD-quality audio and video within reach to all Internet users.

The next generation of videoconferencing software will include a number of compelling features such as on-the-fly data encryption and multiproduct interoperability through adoption of the H.320 international standard. Many LAN/WAN-based videoconferencing systems already adhere to international standards, but not Internet-based products.

Some industry analysts are predicting a convergence of the phone system with the Internet in the not too distant future, with distance and associated charges no longer being a factor in video- and voice-based communications. This will be replaced by a very reasonable flat monthly rate that is much less than the average long-distance bill right now. Already, carriers such as AT&T, Sprint, and MCI are offering consolidated bill-

---

* Asymmetric Digital Subscriber Line (ADSL) is an affordable local loop upgrade technology that allows telephone companies to offer transmission rates of more than 6 Mbps over existing twisted-pair copper wiring—1.544 to 6.144 Mbps downstream (to the customer) and 16 to 640 Kbps upstream (to the telephone company). This is enough bandwidth to support multimedia—video, audio, graphics, and text—to homes and businesses. This technology will also make Web applications work almost as fast as the applications that run on your computer.

ing for a variety of voice and data services, allowing customers to reap increased savings from discounts on the total call volume. This model could be a precursor to future communications services offered over the Internet. At this writing, whether telephone companies participate in or fight the trend toward Internet-based communications is still an open question.

Sending Mail Over the Internet

Introduction

Electronic mail enhances personal and business transactions by replacing paper and manual delivery with faster, more reliable computer communications. With information stored in digital form and conveyed electronically between computers over the Internet, it can be easily edited and integrated into other applications. The popularity of e-mail in recent years has paralleled that of the Internet. More mail is now delivered in electronic form over the Internet than is delivered by the U.S. Postal Service. That amounts to a few billion messages a month!

In recent years e-mail has become an indispensable communications tool for home-based businesses, as well as large companies. In fact, without e-mail, a home-based business would find itself at a great competitive disadvantage and would probably not succeed over the long term. The reason is that most other companies understand the benefits of e-mail and have invested a lot of money in its implementation. If you—as a self-employed entrepreneur working out of a home-based office, for example—want to do business with larger companies, you must use the same communications tools they use.

If you do not have e-mail, you force prospective clients to revert to paper handling and manual delivery, and endure the consequent delay. In this information-intensive age, where the speed of information delivery often spells the difference between success and failure, you cannot expect clients to change their business processes to accommodate your way of doing things. Instead, you must demonstrate to them that you are capable of facilitating the business arrangement rather than posing an undue burden. One of the ways to do this is by using e-mail effectively. Automating information delivery and processing with electronic mail can dramatically reduce the cost of doing business, since the manual tasks of sorting, matching, filing, reconciling, and mailing paper files are virtually eliminated. There are also attendant cost savings on overnight delivery services, supplies, file storage, and clerical personnel.

If you have a home page set up on the Web, you can mail-enable several types of electronic forms from simple guest book and feedback forms to sophisticated purchase order and subscription forms. The forms are built using HyperText Markup Language (HTML) tags and sent via special Common Gateway Interface (CGI) programs written in such languages as Perl or tcl. Not only can you have the completed forms sent back to you, but when retrieved by your e-mail program, they can be sorted automatically upon receipt into appropriate folders and distribution lists. This mechanism is set up using the e-mail program's filtering options and it eases the chore of keeping your mail organized.

Of course, e-mail is not just for business use. It can just as easily be used for personal communications between friends, relatives, and acquaintances located anywhere in the world—and without incurring long-distance phone charges and postage costs. The added advantage is that message delivery takes only minutes, instead of days or weeks, as is typical with postal services in many countries. You do not even need an Internet access account to send and receive e-mail over the Internet. There are free software programs such as Juno and Freemark (discussed later) that specialize in providing this kind of service, the cost of which is covered by advertisers.

Through mail gateways, you can send e-mail to people who may subscribe to some other type of service such as CompuServe, America Online, and Microsoft Network (MSN). Some gateway services, such as RadioMail, can even transport your messages from the Internet to wireless devices such as personal digital assistants (PDAs) and portable computers equipped with radio modems. And with more paging services supporting short-text messaging, you can even send messages over the Internet to alphanumeric pagers (see Chap. 7, "Paging Over the Internet").

The delivery process is entirely transparent, meaning that you do not have to do anything special to send messages to users of these types of services. All you need is their e-mail address. It is the e-mail address that determines how the message is routed. The servers and gateways on the Internet take care of message routing and delivery. If e-mail cannot be delivered because of some problem on the Internet, you will get back a message that not only explains the reason, but estimates the time of delivery.

RadioMail, for example, provides a gateway service that lets e-mail users on virtually any type of service, including the Internet, send messages to any other user via wireless networks. A message addressed as username@radiomail.net is routed to a RadioMail gateway on the Internet. From the gateway, the message is sent over the RAM Mobile Data or ARDIS radio frequency networks to RadioMail's control center where all the translations and coding needed to deliver the message are completed. RadioMail users just turn on their radio modem to receive incoming messages automatically. There is no need to log on or enter a special access code. Delivery is guaranteed; if the radio modem or computing device is turned off, incoming messages are stored until they are turned back on.

The following list provides some examples of how e-mail addresses are formatted to reach users on different services.

Service	*E-mail address*
AT&T Mail	username@attmail.com
America Online	username@aol.com
CompuServe	00000.0000@compuserve.com (change commas to periods)
IBM Mail	username@ibmmail.com
MCI Mail	000.0000@mcimail.com
Microsoft Network	username@msn.com
Prodigy	username@prodigy.com
RadioMail	username@radiomail.net

This brings up another advantage of e-mail: it uses a method of delivery called *store-and-forward*. This enables you to send messages to anyone, anywhere—whether or not that person happens to be online at the moment your message is sent. The next time the person turns on the computer and connects to the Internet service, that user will be notified that new e-mail is waiting. This is a more convenient delivery mechanism than conventional facsimile, which requires that you establish a connection with the remote device before the document can be sent. If the remote device is busy or out of paper or ink, document delivery is delayed until the line becomes free or someone notices the need for more paper or ink.

System Requirements

E-mail is one of the oldest Internet applications, so there are dozens of programs for every type of machine and operating system—even some for the newer handheld personal digital assistants (PDAs). The evolving PDA market has created the need for Internet e-mail solutions for the increasing number of mobile users who rely on electronic messaging for their daily business and personal needs. One of the most popular PDAs is Apple Computer's MessagePad, which uses the Newton operating system. Among the first e-mail programs for the Newton operating system is QUALCOMM's Eudora Light, which also comes in versions for Windows and Macintosh systems.

E-mail software generally poses no unusual requirements in the way of computer resources. In the case of 16-bit Windows 3.x products, a 80386-based PC will do, whereas 32-bit Windows 95 products require at least a

80486 and run better on a Pentium-based machine. Memory requirements are modest: 4 MB RAM or higher for Windows 3.x, but 8 MB or higher is recommended, particularly for Windows 95 users.

E-mail programs do not take up much disk space. In compressed form for download, feature-rich products such as Eudora Pro, Pegasus Mail, and Pronto Mail take between 1 and 3 megabytes (MB); after installation they require between 6 and 7 MB of disk space. Integrated products such as Netscape Navigator 3.0, which include e-mail (i.e., Netscape Mail) and newsreader (i.e., Netscape News) capabilities, in addition to a sophisticated Web browser and CoolTalk plug-ins for text chat, audio conferencing, and whiteboarding, can consume 10 MB of disk space upon installation.

The choice of monitor is usually determined by other applications. A VGA display with 256 colors set for a resolution of 640 × 480 pixels is all you really need for e-mail. The exception is Graphic E Software's Graphic E-Mail for Windows, which requires Super VGA (and 8 MB of RAM). This software lets you encapsulate e-mail messages (Fig. 4.1) within a custom graphic created with an add-on utility. Alternatively, registered users can download a number of ready-to-use templates. Within the graphic is a scroll bar that lets you read long messages. You can also add sound effects—including MIDI music files—and animations.

Installation

To get your e-mail program to work, you must enter some information in its configuration or setup dialog box, including host login, password, and

Figure 4.1
A *Star Trek* motif designed by Rob Schueler is used as a wrapper for this e-mail message.

mail server (Fig. 4.2). The *host login* is the ID you use to access the host. If your e-mail address is maryjane@domainname.com, your host login is probably maryjane. Next, enter the password given to you by your Internet service provider. If you do not enter a password in this field, you will be prompted for it every time you want to use the program for sending or receiving mail.

The final item to enter in this dialog box is the mail server, which is sometimes called the *POP3 server.* This is the Post Office Protocol (version 3), an Internet standard which defines a mechanism for accessing a mailbox located on a remote host machine. This protocol is necessary because your PC will usually not be running all the time and consequently will not always be available to receive mail directly from the Internet. A larger machine, such as a UNIX system, typically will be online all the time, and so is better suited to receiving your e-mail for you. POP3 only defines the retrieval of new mail from the UNIX system, whereas SMTP (Simple Mail Transfer Protocol) is used for sending e-mail.

The mail server is maintained by your Internet service provider and is the place where all incoming e-mail is kept until you download it. You can find out the name of your mail server simply by asking your Internet service provider or system administrator. It might be something as simple as:

mail.domainname.com or domainname.com

You will also need to enter additional information, such as your name and title, the SMTP server, and your return address (Fig. 4.3). In the name and title field, you would enter your real name as you want it to appear in the e-mail messages you send. You can also add your title if you wish. Next, enter your SMTP server. This is maintained by your Internet service provider and is the place where you send all your outgoing messages for

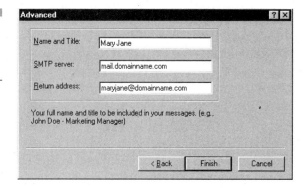

Figure 4.3
Pronto Mail's
advanced settings
dialog box.

delivery over the Internet. You can ask your Internet service provider or system administrator for the name of your SMTP server. More often than not, it has the same name as the mail server. The last item to enter is your return e-mail address. This will show up on the e-mail messages you send, making it easier for others to reply.

If you are using a Windows 95 version of the e-mail program, a short-cut icon may be created for you and placed in your desktop automatically at the time of installation. When you establish a connection to the Internet, you can then click on the shortcut icon to quickly open the e-mail program. This method of access lets you get to your mail faster, without going through the awkward cascading menus of Windows 95.

Most e-mail programs allow you to compose messages offline. To send or receive messages, you must have the Internet connection established. Run whatever Internet connection software you normally use for accessing the Internet (dialers, Trumpet, or other Winsock software) and establish your connection. If you get an error message when trying to send or receive mail, it is probably because you do not have an Internet connection established. Other sources of errors have to do with configuring the e-mail program to recognize your mail server (i.e., POP3 server) or SMTP server.

E-mail Software

All e-mail products offer the same basic functions. Menu options guide the user through the process of creating, editing, sending, receiving, reading, sorting, storing, and printing messages. They differ in terms of advanced features, mail processing capabilities, level of customization, and integration with other applications.

Many Windows-based e-mail programs make extensive use of *drag and drop*—selecting an item by holding down the left mouse button, dragging it to where you want to use it, then dropping it by releasing the mouse button. With Pegasus Mail, for example, you can perform the following actions via drag and drop to expedite your e-mail chores:

■ You can drag address book entries to any address or text field in the program.

■ You can move messages from any folder window to any other folder window.

■ You can copy or move an address book entry from any address book to any other address book.

■ You can drag messages from any folder to any address book and an address book entry will be created based on the information in the message.

■ You can drag a message from any folder window to any message editor field and the contents of that message will be pasted into the message you are editing.

■ You can highlight text in any message you are reading and drag it to a message you are editing.

■ You can drag any file from the File Manager and drop it onto the minimized Pegasus Mail icon to add the file as an attachment to the message you are preparing or begin a new message with the file added as an attachment.

A useful feature of many e-mail programs is the ability to send documents as attachments to messages. However, when you send a document in this manner, it is not always possible to transmit it in its native format. Sometimes it is necessary to package the attachment so the mail transport system or the recipient's e-mail program can understand it. For example, if you are trying to send a document done in Microsoft Word 6.0 to a LAN user who will view it in cc:Mail, the attachment will not open unless it arrives UNIX-to-UNIX Encoded (UUE). When the UU-encoded document reaches the cc:Mail server, it is automatically decoded into Word 6.0.

In addition to UU, other common encoding/decoding formats for attachments are XX, BinHex, and MIME. XX is similar to UU, but uses a different character set than UU so that character-set translations will work better across multiple types of systems—between IBM's mainframe-oriented EBCDIC and ASCII, for example. BinHex is used in the Macintosh world and includes error checking and compression. MIME (Multipurpose Internet Mail Extensions) allows different mail applications to exchange a variety of types of information, including attachments in GIF and TIFF graphics formats.

Some e-mail programs perform encoding/decoding automatically, requiring you to intervene only in very special cases. Some e-mail programs do not include the capability to encode/decode attachments at all. In such cases, you will have to rely on third-party shareware programs such as Sabasoft's Information Transfer Professional (Xferpro), which can automatically handle the decoding of many common formats. When decoding, the procedure is to detect the file format and decode as appropriate. Due to the prolific number of formats, however, it is possible that Xferpro can be confused by an unsupported format. To assist in decoding, you can request a specific decode file format (Fig. 4.4). In addition, if the encoded input is spread across multiple files—as it might be in multipart attachments—Xferpro will treat these files as one large file to be decoded.

Filtering allows you to automate the processing of mail when messages matching particular conditions are met. For example, you might want to have all "confirmation of delivery" or "confirmation of reading" replies appear in a special folder called Confirmations. This would keep your other folders from getting clogged with these routine messages, especially when they are the result of mass mailings. At the same time, you can find this information more easily when it is placed in its own folder. To do this,

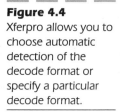

Figure 4.4
Xferpro allows you to choose automatic detection of the decode format or specify a particular decode format.

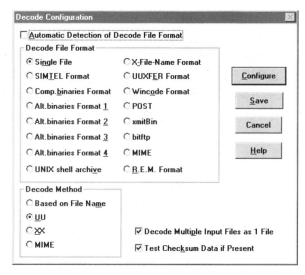

you set up a folder named Confirmations. Then you specify the condition that the filter will test against. In this case, the test would be the word *receipt* or the text string *rcpt* either of which usually would be found in the Subject: field of these types of messages (Fig. 4.5).

Among the other actions that can be taken in response to rule triggers defined in Pegasus Mail are the following:

- *Add user to list.*   Adds the e-mail address of the sender to a distribution list, with no duplication
- *Copy.*   Copies the e-mail message to another folder
- *Delete.*   Deletes the e-mail message from the folder
- *Extract.*   Saves the text of one or more messages to files on your hard disk
- *Forward.*   Forwards messages to another e-mail address
- *Highlight.*   Applies a color to the trigger text
- *Print.*   Prints the message to the default printer
- *Remove user from list.*   Removes the e-mail address of the sender from a distribution list
- *Send text file.*   Sends a text file in response to an e-mail message
- *Send binary file.*   Sends a binary file in response to an e-mail message

Multiple rules can be defined with the same trigger conditions to have multiple actions applied to the same message. The actions will be applied

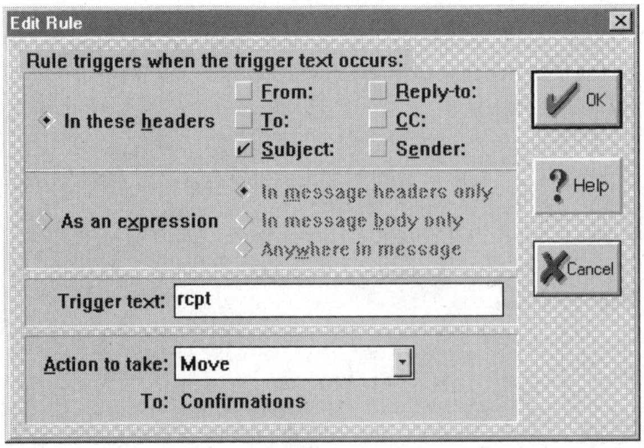

Figure 4.5
In Pegasus Mail, a rule specifies that the Subject: field of all incoming messages will be checked for the text string *rcpt,* indicating a confirmation of delivery or a confirmation of reading. Any matching messages will be automatically moved to the Confirmations folder when mail is downloaded from the server.

in the order in which they appear in the actions list. Rule processing continues until all rules have been applied or until the message is moved to a folder or deleted as the result of a rule.

Features

E-mail programs differ in terms of performance and features. After trying out a few e-mail programs to see how well they perform, you can focus in on comparing features to make a final selection. What follows is a list of features you may want to consider:

- *Autoforwarding.* Redirects your mail to another account. This is especially useful when you are on vacation and someone else must handle your job responsibilities. You also have the option of receiving a copy of your e-mail, even as it is autoforwarded to someone else.

- *Automated responses.* Lets you automate replies to messages, either selectively or globally. For every incoming message, you can have a reply sent indicating that you are on vacation, for example, and that you will not be checking e-mail until you return. Alternatively, you can choose to receive only high-priority messages or messages from specific e-mail addresses.

- *Clickable e-mail addresses.* Automatically opens a message compose window when you click on an e-mail address appearing in the body of the message.

- *Clickable URLs.* Automatically opens your World Wide Web browser and goes to the selected URL (Uniform Resource Locator) appearing in the body of the message. (The mouse pointer changes to a pointing hand when you pass it over an active link.)

- *Color-coded message labeling.* Lets you apply different colors to the parts of a message, such as the body text, URL or e-mail addresses, and reply text.

- *Compression.* After messages are deleted from file folders, the e-mail program performs compression to reclaim disk space.

- *Context-sensitive online help.* Lets you click on items in the user interface to read brief explanations of what the items do.

- *Default routing.* Lets you specify the default address for recipients in the address book who have multiple e-mail addresses.

■ *Distribution lists.* This feature lets you create a list of addresses for mass mailings. This list has its own name and when entered in the To: field, the message and attachment are sent to all addresses on the list.

■ *Draft.* You can save incomplete messages in a file folder to continue working on them at a later time. To make finding the draft messages easier, you can name the file folder Drafts.

■ *E-mail alerts.* You can configure the e-mail program so that when it is running in the background an alert will sound when new mail arrives.

■ *Exceptions.* Allows you to exempt specified addressees from a mailing list, without having to create a new list.

■ *Expanded text.* You can store abbreviations for commonly used text strings which you can expand at any time with a single key stroke.

■ *Filters.* Allows you to do such things as automatically identify and file your mail by having messages sorted into mailboxes or distribution lists, forwarding mail to other locations or persons, having automated replies sent, and alerting you when important messages arrive.

■ *Graphic wrapper.* Allows you to create and send a graphic image which acts as a message wrapper. It can contain an integral scroll bar to aid message viewing.

■ *Image display.* Allows you to include an image of yourself in your e-mail messages.

■ *Incremental mailings.* Allows you to break up very large lists to expedite mailing.

■ *List suppression.* In distribution lists, instead of having the e-mail address of every recipient listed in the To: field, you can suppress them by substituting something else, such as "Distribution" <yourname@domainname.com>, where "Distribution" (including quotes) is the name of the mailing list and <yourname@domainname.com> is your e-mail address (including angle brackets).

■ *Long filenames.* Allows the use of filenames longer than the eight-character DOS limit (Windows 95 mail clients only).

■ *Mailbox management.* You can create your own mail folders to arrange incoming messages by date, time, author, and subject. (See also the preceding description of filters.)

■ *MAPI support.* A MAPI (Mail Applications Programming Interface) utility that lets you send e-mail with attached files directly from your MAPI-compliant applications—word processors, spreadsheets, graphics, and other applications.

■ *Multiple accounts management.* You can manage multiple e-mail accounts and switch between them via hot keys or the point-and-click method.

■ *Nested mailing lists.* Allows you to embed other mailing lists (child lists) within another mailing list (parent list).

■ *Nicknames.* Also called *aliases,* these allow you to keep a list of individuals you send messages to frequently, equating their real name to an e-mail address. When you want to send messages to these individuals, you can enter their real name in the To: field instead of an e-mail address, which might be difficult to remember or inconvenient to look up in an address book.

■ *Poll time.* You can control how frequently the e-mail program scans the server for new mail while the program is open and connected to the Internet.

■ *Reminder.* Also called a *tickler,* this allows you to send messages to yourself with a specified delivery date so you do not forget important occasions, deadlines, or events.

■ *Remove expired messages.* Automatically removes messages from the server if the message is *x* days old and has already been retrieved from the server.

■ *Scheduled delivery.* You can compose messages offline and save them in an outbox and then have them automatically sent all at once at a more convenient time.

■ *Signature.* Allows you to create a standard signature or *sign-off* that is automatically placed at the end of your e-mail messages. This signature might include your name, title, organization, phone and fax numbers, and the URL of your Web page. Some e-mail programs allow you to create multiple signatures—for personal or professional use, for example. You might also use different signatures, depending on whether the e-mail is for local, Internet, or MHS (Novell's messaging system) delivery.

■ *Spell checker.* Checks the spelling of text in your e-mail messages. Some e-mail programs allow you to ignore original text in your replies, which saves time by not checking text that came from a previous message.

■ *Stylized text capabilities.* Lets you format messages using various fonts, colors, and sizes.

■ *UNIX-to-UNIX encode/decode.* Some e-mail products include the capability to automatically UU encode/decode files without resorting to external programs.

- *Voice recording.* Allows you to record an audio message and send it as an attachment.

Security

E-mail sent over the Internet is less secure than a postcard sent through the mail. As your e-mail passes through the various servers on the Internet on the way to its destination, it is exposed to anyone who wants to read it. All that needs to be done is for system administrators or technicians to collect the packets that pass through their site's equipment (such as a router) by attaching a *traffic analyzer.* This device, which is designed for troubleshooting problems on the network, has the capability to record and decode packets all the way up to the application layer, meaning that the text of your messages can be played back and read with very little trouble.

However, if you encrypt your messages, you can protect your information from prying eyes. You can use PGP (Pretty Good Privacy) software for this purpose, which is downloadable from MIT's Web site at:

http://web.mit.edu/network/pgp-form.html

As noted in Chap. 2, PGP is distributed by MIT with the understanding that you are a U.S. citizen located in the United States or a Canadian citizen located in Canada. PGP generates two keys that belong uniquely to you. One PGP key is secret and stays in your computer. The other key is public and is given out to people you want to communicate with. The key you distribute enables others to decrypt your messages back into their original form.

There is encryption software that is much easier to use than PGP. One such product comes from Scrambler Technologies. Its Scrambler is a Windows program that incorporates sophisticated Data Encryption Standard (DES) technology to scramble text and graphic files.

Recipients of the information use Scrambler, together with an agreed-upon keyword, to decrypt the unintelligible material back to its original form. The result is complete security and privacy for information transmitted via the Internet, stored on hard drives and disks, or sent through internal e-mail systems.

To scramble a document within a word processing program, for example, you simply select and block the desired text, and click Scramble in the Windows system menu, and choose a key word. The desired information is then encrypted and ready for secure transmission or storage.

You can obtain more information about Scrambler, including pricing, from the company's Web page at

http://www.aescon.com/scrambler/

Some e-mail products, such as Graphic E-Mail, include a security mode which provides simple text encryption. However, such proprietary schemes can only be used if both the sender and receiver use the same e-mail software.

Mail-Enabled Web Pages

As noted earlier, an important aspect of operating a home page on the Web is being able to collect information from visitors via forms. For example, companies might post product catalogs and use forms to collect purchase information, including credit card numbers. A publisher might post samples of a newsletter with the idea of inviting visitors to fill out an order form for an annual subscription.

Even if you do not use your Web page for commercial purposes, you can benefit from the use of forms. You might post a survey form asking visitors what they liked or disliked about your Web page, so you can tailor content more appropriately. You can create a form that asks visitors for their name and e-mail address if they wish to be notified of changes to your Web page. And, of course, there is the standard guest book form, which invites visitors to sign in when they access your Web page.

Whatever the purpose, forms are used to collect vital information. The forms themselves are created using the HyperText Markup Language (HTML). The information entered into these forms is processed and sent to you via e-mail by Common Gateway Interface (CGI) scripts. When the forms data arrives via e-mail, you can use the filtering capabilities of your mail program to completely automate the handling of this information. What follows is a complete description of how to do all this, starting with the creation of a simple form in HTML and writing a CGI script to process and send the data, and then setting the e-mail program's filters to automate the handling of the information.

For this demonstration we will create an Email Notification Form that can be accessed as a menu item on your home page. With this form, visitors can request to be notified via e-mail whenever there is a change to your Web pages. This saves them the time and trouble of continually checking back to see if anything has changed. By filling out the form, the user can indicate whether he or she wishes to subscribe or unsubscribe to your Email Notification Service. If subscribe is selected, that person is added to the distribution list. If unsubscribe is selected, that person is removed from the distribution list. In either case, the user will get an on-screen acknowledgment message, indicating that the appropriate action has been taken.

What follows is a very simple method of implementing this scenario. It requires an HTML form, a CGI script written in Perl, and a mail program such as Pegasus Mail that can apply the various filters we will need.

The HTML Form

The first item you will need is an appropriate HTML form.* The follow-ing HTML code will get you started, but you can customize it any way you like. Among the important features that are demonstrated here is the use of multiple Submit buttons: one labeled Subscribe and the other labeled Unsubscribe. These buttons allow users to notify you of whether they would like to get on or off your mailing list. These labels will be used later in the filtering process that automatically adds or removes names from the distribution list. Other items that merit further discus-sion are highlighted in bold.

```
<HTML>
<HEAD>
<TITLE>Email Notification Service</TITLE>
</HEAD>
<BODY>
<P>
<CENTER><H2>Email Notification Service</H2></CENTER>
<P>
<HR=5>
<P>
<CENTER>To be notified by e-mail whenever a new item appears on
this Web page, just fill in this form.</CENTER>
<P>
<FORM METHOD="POST" ACTION="/nmullerbin/notify.pl">
<B>Name:</B><BR>
<INPUT TYPE="text" NAME="name" size=40>
<P>
<B>Organization:</B><BR>
<INPUT type="text" NAME="organization" size=40>
<P>
<B>Email Address:</B><BR>
<INPUT type="text" NAME="e-mail" size=40>
```

---

* This discussion assumes you are familiar with the use of HTML tags. If not, my book *The Webmaster's Guide to HTML,* published by McGraw-Hill (1995), explains these tags in more detail, including their use in the creation and processing of many types of forms.

```
<P>
<CENTER>
<INPUT type=SUBMIT VALUE="Subscribe" name="Subscribe">
<INPUT type=Reset value="Clear">
<INPUT type=SUBMIT VALUE="Unsubscribe" name="Unsubscribe">
</CENTER>
</FORM>
<P>
<HR>
<P>
<A HREF="index.html"><IMG SRC="back.gif" ALIGN=MIDDLE></A><EM> Go
back to Main Menu</EM></P>
</BODY>
</HTML>
```

The results of this HTML coding are shown in Fig. 4.6.

All forms must be encapsulated within the <FORM> and </FORM> tags. You will notice that the first of these tags makes reference to the attributes METHOD, POST, and ACTION. These refer to how forms are processed by the CGI script. METHOD defines the technique used to submit the form information to the server. METHOD can have either of two arguments:

Figure 4.6

The HTML tags are used to create this form for the Email Notification Service, as rendered in Netscape Navigator.

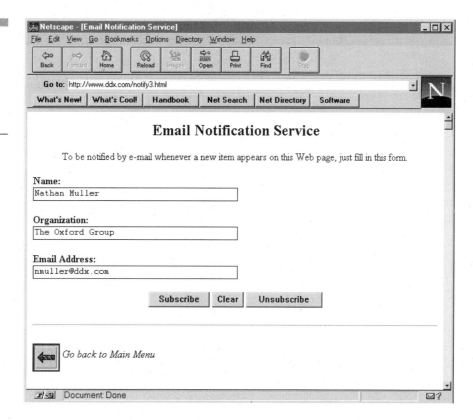

- *GET (the default).* The form contents are appended to the URL.

- *POST (recommended).* The form contents are sent to the server in the body of the message. To take advantage of POST, your Internet service provider must support the POST method, and most of them do.

ACTION specifies the URL of the Perl program to which the form inputs will be sent. The program should reside in the cgi-bin directory at your Internet service provider's Web server. In this example, the location and name of the Perl file is simply /nmullerbin/notify.pl. The reference nmullerbin is a *bin alias* that is set up at the Web server, while notify.pl is the name of the Perl script that supports this form (discussed later). The bin alias is merely shorthand for the location of the script on the server. On a shared Web server, directory/subdirectory names can be quite long. For example, using nmullerbin saves me from having to type the following whenever I want to create a new form:

.../home/users/nmuller/public_html/cgi-bin/

If you create a lot of forms, you will appreciate having your own bin alias. It takes only a few seconds for your Internet service provider or system administrator to set you up with one.

The next items in the HTML-coded form that merit explanation are the input fields for the name, organization, and e-mail address. In this form, the INPUT tag is used to indicate a single-line text field. The NAME attribute is used to identify the particular field when the form's contents are processed by the Perl script. The SIZE attribute specifies the visible width of the field in characters.

After the user fills out the evaluation form, he or she can submit it by clicking one of the Send buttons or reset it by clicking the Clear button. These also are defined with the INPUT tag. The two possible VALUE attributes are submit (for Subscribe and Unsubscribe) and reset (for Clear). The labels on these buttons can be changed to anything you want. When rendered, the size of the button will expand (or contract) to accommodate the length of the label.

The next step is to equate these fields to the Perl script so that the form's contents can be routed and processed properly. For this, we will create a CGI script in Perl and name it *notify.pl,* which is already referenced in the first FORM tag of the HTML document.

The Perl Script

A CGI script is used to process the inputs from HTML forms. For this example, we are using a very simple script written in Perl. You can use the script as is or customize it any way you like.* Items that merit further discussion are highlighted in bold.

```perl
#!/usr/local/bin/perl

$sendmail = "/usr/lib/sendmail";
$to = 'yourname@domainname.com';

print "Content-type: text/html\n\n";

read(STDIN, $buffer, $ENV{'CONTENT_LENGTH'});

@pairs = split(/&/, $buffer);
foreach $pair (@pairs)
{
   ($name, $value) = split(/=/, $pair);
   $value =~ tr/+/ /;
   $value=~ s/%([a-fA-F0-9][a-fA-F0-9])/pack("C", hex($1))/eg;
   $value =~ s/~!/~!/g;
   $FORM{$name} = $value;
   substr($FORM{'subscribe'},$[,1)=~tr/A-Z/a-z/;
   substr($FORM{'unsubscribe'},$[,1)=~tr/A-Z/a-z/;
}

if ($FORM{'e-mail'} !~/.+\@[\w.]+/){
print "<H2>This form is not complete.</H2><P>\n";
print "Please go back and provide an e-mail address.<P><HR><P>\n";
print "<A HREF=\"http://www.iquest.com/notify.html\"><IMG
SRC=\"http://www.iquest.com/button.gif\" ALIGN=MIDDLE></A><EM>
Return to the Email Notification Form\n";
exit;
}

open (SMAIL, "|$sendmail $to") || die "Can't open $sendmail!\n";
print SMAIL "To: yourname@domainname.com (Your Name)\n";
print SMAIL "From: $FORM{'e-mail'} ($FORM{'name'})\n";
print SMAIL "Organization: $FORM{'organization'}\n";
print SMAIL "Subject: Email Notification Service\n\n";
print SMAIL "-------------------------------------------------\n";
print SMAIL "I would like to $FORM{'subscribe'}$FORM{'unsubscribe'}
to your Email
Notification Service. \n";
print SMAIL "-------------------------------------------------\n";
close (SMAIL);

print "<Head><Title>Email Notification Service</Title></Head>";
print "<P>\n";
print "<H2>Thank you.</H2></Body><P>";
print "<H3>Your request to $FORM{'subscribe'}$FORM{'unsubscribe'}
to the Email Notification Service will be processed immediately.
</H3><P>\n";
```

---

* This discussion assumes no familiarity with Perl. My book *The Webmaster's Guide to HTML,* published by McGraw-Hill (1995), provides more detail, including how to write Perl scripts to process many types of forms, including those that handle credit card transactions.

```
print "<HR><P>\n";
print "<A HREF=\"http://www.iquest.com/index.html\"><IMG
SRC=\"http://www.iquest.com/button.gif\" ALIGN=MIDDLE></A><EM>
Return to
Main Menu </EM>\n";
```

The Perl script begins with the following line:

```
#!/usr/local/bin/perl
```

This line tells the server to run the script through the Perl interpreter. The program will not work without this line. In addition, this line must be followed by a blank line.

The next line,

```
$to = 'yourname@domainname.com';
```

is your e-mail address, indicating where the form will be delivered.

The next section of the script specifies the content type; in this case, text and/or HTML-coded text. After that, the next lines in the program establish the relationships of name and value pairs.

The next significant lines in this Perl script are

```
substr($FORM{'subscribe'}, $[, 1)=~ tr/A-Z/a-z/;
substr($FORM{'unsubscribe'}, $[, 1) =~ tr/A-Z/a-z/;
```

These lines perform pattern matching on the words Subscribe and Unsubscribe which are the labels on the form's submit buttons. Specifically, these lines are used to change the letters *S* and *U* to lowercase so they can be output correctly in midsentence.

To prevent users from submitting forms without an e-mail address and possibly cluttering up your mailbox with useless information, you can refuse to accept the form unless the e-mail field is filled in (Fig. 4.7). This is accomplished with the following lines in the Perl script, which performs pattern matching to detect the format of a validly constructed e-mail address:

```
if ($FORM{'e-mail'} !~/.+\@[\w.]+/) {
print "<H2>This form is not complete.</H2><P>\n";
print "Please go back and provide an e-mail address.<P><HR><P>\n";
print "<A HREF=\"http://www.iquest.com/notify.html\"><IMG
SRC=\"http://www.iquest.com/button.gif\" ALIGN=MIDDLE></A><EM>
Return to
the Email Notification Form\n";
exit;
}
```

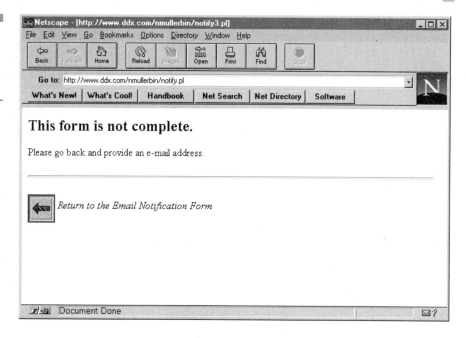

The part of the Perl script that actually causes the HTML form's contents to be output as e-mail, and in the format you specify, is encapsulated with the open SMAIL and close SMAIL commands. Wherever $FORM appears, with a label in curly braces, the HTML form's inputs will be printed.

For example, $FORM {'name'} refers back to the field in the HTML-coded form that asked for the user's name. That field was given the input name of "name", which is referenced in the Perl script as 'name'. The only difference is that in the HTML version, double quote marks are used and in the Perl version, single quote marks are used. The same convention is used for the organization and e-mail address fields. Wherever $FORM {'organization'} and $FORM{'e-mail'} appears in the Perl script, the user's organization and e-mail address will be output.

The same is true of $FORM{'subscribe'}$FORM{'unsubscribe'}. Notice in the Perl script how the references to these fields run together without a space between them. This is because the user can only click on one button—Subscribe or Unsubscribe—to send the form. Regardless of the choice, it will show up in midsentence with correct capitalization.

Initializing the Script

Before you can use the Perl script, you must put it into a cgi-bin subdirectory on the server. If you do not already have a cgi-bin directory, you must either establish a Telnet session with the server to create one or request that the system administrator create it for you. The HTML-coded forms can be located in any directory, preferably with your other HTML documents. Typically, HTML documents and programs are loaded into their respective directories via FTP. Once the Perl scripts are placed in the cgi-bin subdirectory in ASCII form (never save them in binary form), they must be initialized so they will be executable. You can have the system administrator do this for you, or you can do it yourself via a Telnet session. After loading the Perl script to the cgi-bin subdirectory via FTP, just access the directory via Telnet and perform the following function (assuming a UNIX server) the first time the Perl script is loaded:

```
chmod a+rx notify.pl
```

This determines who gets to use the file for processing forms. Essentially, *chmod* means change mode, which makes the script executable and sets the permissions; specifically, this allows notify.pl to read (r) and execute (x) for everyone (a+). You only have to do this procedure the first time you load a Perl script. You can modify and reload a script any number of times via FTP and the permissions will hold unless specifically changed. Any time you modify and reload the Perl script via FTP, it must be as an ASCII file and not binary. Uploading the file as binary will introduce hard returns into the Perl script which will render the script unusable.

E-Mail Output

Assuming that the user decided to subscribe to your Email Notification Service, the Perl script will send you an e-mail message in the following format:

```
Date: Mon, 23 Sep 1996 12:27:22 -0500
To: yourname@ddx.com (Your Name)
From: username@ddx.com (User Name)
Organization: The Oxford Group
Subject: Email Notification Service

------------------------------------------------------------
I would like to subscribe to your Email Notification Service.
------------------------------------------------------------
```

On-Screen Acknowledgment

The last lines in the Perl script, which begin with the print command, will cause an appropriate acknowledgment message (Fig. 4.8) to be dis-

Figure 4.8
The acknowledgment
message as rendered
in Netscape Navi-
gator.

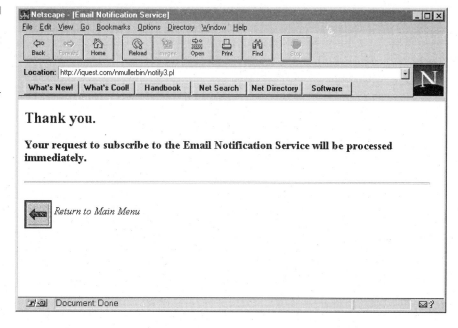

Figure 4.8
The acknowledgment
message as rendered
in Netscape Navi-
gator.

played to the user when either the Subscribe or Unsubscribe button is clicked. This lets the user know that the request is being handled, rather than leaving him or her guessing about whether the form worked.

Applying Filters

As users subscribe to your Email Notification Service, you can set up the filtering capabilities of your e-mail program to redirect these e-mail messages to a separate mail folder based on the contents of the Subject: field, which in this case is Email Notification Service. Depending on the e-mail program, you can even have the senders' names and e-mail addresses automatically added to a distribution list. At a later time, when you want to announce that your Web page has added some new content, you can send a message to all the addressees on the distribution list. If any addressee uses the HTML form on your Web page to unsubscribe from your mailing list, you can set up the filters to automatically remove their name and address from the list when the message is received. The filter does this by checking for the word unsubscribe in the body of the message.

Keep in mind that not all e-mail programs offer filtering. Those that do, such as Eudora Pro and Pronto, offer very basic filtering capabilities, while Pegasus offers advanced filtering capabilities.

E-mail Services

A relatively new way of sending and receiving e-mail over the Internet made its debut in 1996. It is distinguished from conventional e-mail products by the fact that it does not require you to connect to the Internet to send or receive e-mail. You simply dial a local access number to connect to a gateway that actually sends and receives e-mail over the Internet. Juno and Freemark are examples of this type of service. These services are supported by advertisers, enabling them to be offered free to users (Fig. 4.9). All you need to use the service is the software and a modem-equipped computer. In the case of Juno, you are encouraged to copy the software and distribute it to friends, relatives, and coworkers.

Part of Juno's configuration process involves filling out an 18-question member profile describing your interests, hobbies, activities, and other characteristics. Your answers are intended to help Juno understand what kind of information, products, and services you might be interested in. This enables Juno to be selective in choosing the advertisements it displays. If in the future any of this data changes, you are supposed to update the member profile. Juno promises that it will not send you more than one or two direct e-mail messages in any given session, and in most will send you none. However, Juno's sponsors may sometimes send you surveys about their current and future products.

Figure 4.9
Juno's graphical user interface. Advertising appears across the top. You can even click on the advertisements for more information.

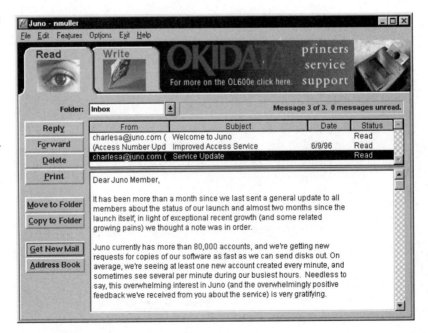

Juno offers basic e-mail functions, including the ability to file messages in folders and an address book that allows you to store commonly used e-mail addresses and to insert them easily into your messages. You may also create nicknames, or aliases, for these addresses, so that instead of having to remember and type in an address such as charlietyke12345@mail. address.comp.com, you can just type a simple alias such as Charlie. You can also create mailing lists so that you can send the same message to several people at once. The spell checker not only identifies misspelled words, but offers replacement suggestions.

The service has some limitations. It is only available for use within the United States, although you can send and receive e-mail internationally. The current version of Juno does not enable you to attach nontext files to a message. Juno cannot be run on a LAN, but the company may issue a Pro version of its software in the future which would provide this capability. Finally, Juno does not provide you with direct access to the Internet, so you can't use it for browsing the Web or Usenet newsgroups or to establish Telnet sessions with remote computers.

Flowmark's service offers comparable features. Both services offer basic features for users who do not use e-mail frequently enough to pay a monthly fee for an Internet access account or who have no need for advanced features. They also make a good backup for e-mail when your Internet service provider is offline due to a system or link failure. You can also use these services to stay in touch when you are traveling. Since you can dial a local number to access Juno, you avoid long-distance charges.

List Servers

In your travels around the Internet, you may come across references to list servers, mailing lists, and majordomo. These terms mean pretty much the same thing, with *majordomo* being a popular server-based list management system—implemented by a suite of Perl scripts written by Brent Chapman—that automates the operation of large mailing lists. List-Processor is another popular mailing list management system. The mailing lists themselves are devoted to a specific discussion topic and can be public or private, moderated or unmoderated. Basically, these mailing list management systems allow subscribers to talk among themselves via e-mail messages that are posted to the list server. Everyone on the list gets a copy of everyone else's messages and is free to send a reply. Anyone can start a mailing list and invite others to subscribe.

Majordomo automates routine administrative tasks such as requests to subscribe or unsubscribe. If the list is moderated, all subscription requests and messages can be routed to the list owner for approval. Once the list is set up at the server, it can be controlled by simple commands sent by the list owner via e-mail to the server.

You can subscribe (or unsubscribe) to a list and do such things as ask for an introduction to the list, request help, and obtain a list of other subscribers. To subscribe to a list, you just address a message to the majordomo server, typically in the format majordomo@hostname. You type in the word *subscribe* (or *unsubscribe*) in the message body, with no other text. Among the commands, including the dash, that can also be put in the body of the mail message are:

- *help.* Sends one-line summaries of majordomo commands
- *info list.* Sends a welcome message and an introduction to the list from the list owner
- *who list.* Provides a list of subscribers, unless the list is private

If you want to start a mailing list, you must have direct access to the server where the Perl interpreter and majordomo are installed. If you do not have such access, your Internet service provider will usually set up and test the list for you. You will probably have to pay a setup fee and monthly charge for this service. Once majordomo is set up, you can manage the list without any additional involvement by your Internet service provider, since majordomo accepts commands sent to it via e-mail. Even though the Internet service provider may not be involved in the daily operation of your list, the monthly charge is justified because its system resources are being used for the operation of the list.

Mailing lists are useful tools for distributing news and announcements or for discussions where it is desirable that people worldwide be allowed to contribute. Unlike chat rooms (see Chap. 10 on "Participating in Newsgroups"), however, mailing lists are not intended for interactive conversation and they can be a less secure method of communication than ordinary e-mail. The reason is that majordomo was not designed for personal messages, but to facilitate communication among members of a large group. Therefore, security is not a key concern of majordomo.

Online Resources

The following table provides links to the Web pages or FTP sites of the major e-mail software providers. From these links, you can download working or evaluation copies of e-mail programs. You can also obtain

Developer	Product	Web page or FTP site
Attachmate Corp.	Emissary, IRMA for Internet	http://www.attachmate.com
Bonzi Software	Voice Mail	http://www.bonzi.com
BrainTree	BrainTree eBase/mail	http://www.braintree.com
CommTouch Software	Pronto Mail	http://www.commtouch.com/
ConnectSoft	E-Mail Connection	http://www.connectsoft.com
Mathew Dredge	E-Mail Thing	http://www.wave.co.nz/pages/mdredge/emthing.html
Freemark Communications	Freemark Mail	http://www.freemark.com:80/freemark.html
Graphic E Software	Graphic E-Mail for Windows	http://www.clever.net/ge/index.html
David Harris	Pegasus Mail	http://www.pegasus.usa.com/welcome.htm
Maurice Jeter and MJS	DurangoMail	http://www.pebbs.com/durango/
Juno Online Services, LP	Juno	http://www.juno.com
Microsoft Corp.	Internet Explorer	http://www.microsoft.com
NetManage	Chameleon Mail	http://www.netmanage.com
Netscape Communications Corp.	Netscape Mail	http://www.netscape.com
QUALCOMM	Eudora Light	http://www.eudora.com/light.html
Quarterdeck Corp.	Quarterdeck Mail, Quarterdeck Message Center	http://www.quarterdeck.com
RIM-Arts	Becky! Internet	http://www.bekkoame.or.jp/~carty/
Software Laboratory	Mail for Windows 95	
University of Washington	Pine	http://washington.edu/pine

NOTE: This information, as well as updates, can be found at http://www.ddx.com/mgh.shtml.

other useful information such as the platforms supported, system requirements, product features, installation instructions, and troubleshooting advice. Since the technology is moving rapidly, you may want to consult these locations periodically for the latest developments.

Conclusion

Not too many years ago, e-mail was considered a fad by most companies. Now there is a great appreciation for e-mail and its role in supporting daily business operations. In fact, the popularity of e-mail is now so great that many companies are having to rethink the capacity of their communications links and systems to accommodate the growing traffic load. Steps are also being taken to minimize unnecessary traffic, such as by storing only one copy of e-mail attachments on a server, rather than allowing the same attachment to be duplicated to all recipients of the message.

The Internet has helped to make e-mail almost as ubiquitous and essential a method of communication as the telephone. Wireless services extend the reach of e-mail even further, since they do not rely as much on the existing wireline infrastructure. In turn, this has given computer users the highest degree of mobility, enabling them to conduct business and stay in touch with coworkers without regard for location, distance from the office, or proximity to a telephone.

We can look forward to increasing integration of audio, image, and video in e-mail programs, making it easier for users to create and preview these files from within the e-mail program itself before transmission. There are even programs that scan and compress documents so they can be sent as faxes over the Internet. Faxing over the Internet is the topic of Chap. 6.

Voice Mail and Video Mail Over the Internet

Introduction

Voice mail and video mail enable users to personalize their messages with actual voice and video recordings, adding interest and excitement to otherwise bland text messages. For years, these technologies have been used by major corporations over private networks. Now they are becoming available to you for use over the Internet. Voice and/or video messages offer the means to extend your personality over the Internet to friends, relatives, coworkers, and clients anywhere in the world. As a result, your messages will have more impact and longer retention.

Of the two, voice mail has been around the longest. Over the years voice mail has become an effective communications tool that can enhance productivity and permit personal mobility without the risk of being out of touch with friends, family, or colleagues. Voice mail provides the convenience of allowing callers to leave messages when you are not available to take calls personally.

Voice mail is implemented in the business environment through a PBX or add-on messaging system. It is also offered as a service by telephone companies and numerous third-party voice messaging companies. Among the advantages of voice mail are

- Ensures that the message is accurate
- Provides the opportunity to leave detailed messages and explanations
- Enables information to be delivered in the caller's own speaking style
- Provides privacy over other message delivery methods such as operators, receptionists, and secretaries.

Video mail usually runs on a company's local area network and adds the advantage of being able to see the person delivering the message. It has been slower to catch on than voice mail, mostly because it requires special equipment—including a video camera, microphone, and speakers for each computer—the use of unfamiliar protocols, and lots of bandwidth. Furthermore, adding bandwidth-intensive video mail could disrupt other applications on the network if it is not managed properly. These problems are being overcome, however, as vendors employ innovative techniques—including prioritization and traffic control schemes—to bring real-time multimedia applications to the desktop.

With the right software, a sound card, and an Internet connection, you can take advantage of voice mail and video mail just as corporate users do—and have fun too. Voice mail and video mail are a great way to communicate, especially if you're a student or in the military and must be

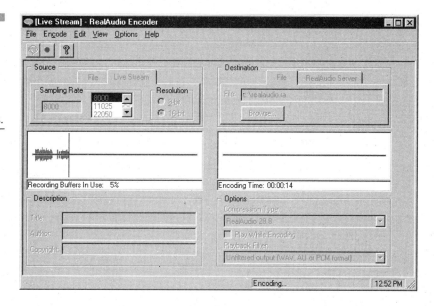

Figure 5.1
Progressive Net-
works' RealAudio
Encoder allows you
to create and edit
voice mail messages
and other sound files.

away from home for long periods of time, perhaps even in another country. Sure, family and friends appreciate letters and postcards, but hearing
your voice and seeing your image is even better.

With regard to voice mail, there are several ways to record and send messages. You can use stand-alone encoder software to create a voice message
and send the file as an attachment using your favorite e-mail software.
Among the most popular voice encoders are Progressive Networks' RealAudio (Fig. 5.1), DSP Group's TrueSpeech, and Voxware's ToolVox (Fig. 5.2).

Figure 5.2
Voxware's ToolVox
Encoder lets you use
or create WAV files
and convert them to
an 8 kHz, 2400 bps
VOX file, which is
more efficient for
transmission over the
Internet.

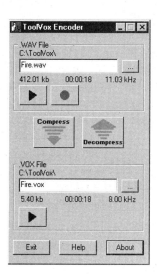

In addition, you can use the Sound Recorder (Fig. 5.3) and Media Player (Fig. 5.4) applications that come with Windows 95 and Windows NT, which are based on technology supplied by TrueSpeech.

You can also use the voice messaging capabilities offered by some Internet phone products, such as NetSpeak's WebPhone, IRIS Systems' IRIS Phone, and VocalTec's Internet Phone. With IRIS Phone, for example, you can send voice mail to a party who is currently unreachable (i.e., not online or already talking with another user) and your message will be delivered as soon as he or she becomes available. VocalTec even offers a free Voice Mail Player (Fig. 5.5), which lets recipients play back voice mail messages if they don't happen to have a copy of Internet Phone (version 4.0). The player is available for download at

http://www.vocaltec/.com

Another solution is to use the voice mail capabilities offered by some e-mail programs such as Pronto or Bonzi. This approach allows you to open the voice encoder from within the e-mail program, review the message, and send it just as you would any other e-mail message. In addition to an Internet phone with a voice mail capability, VocalTec offers a stand-alone voice mail program and player called Internet Voice Mail.

Regardless of the specific approach, in essence, all you're doing is sending a sound (or video) file as a MIME attachment over the Internet. When it arrives at its destination, the file can be opened with any player that supports the same file format.

Figure 5.5
VocalTec's Voice Mail Player lets recipients play back your voice mail messages sent using Internet Phone (version 4.0).

System Requirements

To record voice or video messages, you must have a computer with enough processing power to handle the encoding/decoding algorithms the software vendors use for capturing your voice and compressing the resulting files to a reasonable size. For this to work properly, you should have a 486/33DX computer or better. Some encoding algorithms such as GSM—which is supported by NetSpeak and VocalTec, among others—require a Pentium processor. And if you want to manipulate these files by adding special effects or music from a compact disc (CD) or other input source, a Pentium is a virtual necessity.

You also should take note of your available disk space because audio and video files can be quite large. For example, a one-minute audio file created in the RealAudio format will use about 66 kilobytes (KB) of disk space when recorded at 14.4 Kbps and about 96 KB of disk space when recorded at 28.8 Kbps. The higher recording rate will result in better voice quality but will consume more resources. If you're already running low on disk space, it's a good idea to add another disk drive before getting serious about voice or video mail.

To record and playback audio and video files, you'll also need a sound card that is compatible with your type of computer, speakers (or headset), and a microphone. Of course, to create video mail, you will also need a camera. Fortunately, video cameras for Macintosh and IBM and compatible PCs are very inexpensive, in some cases, priced at less than $100. (More on video mail later.)

Software

To create and play sound files you will need two software components: an encoder and a player. The *encoder* creates compressed sound files, either

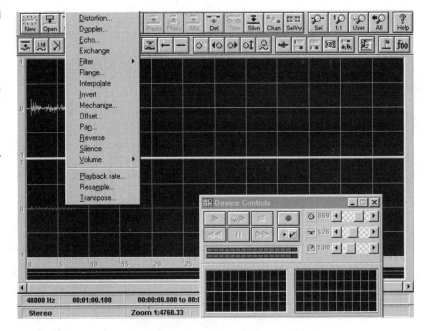

Figure 5.6

Chris S. Craig's shareware program Gold-Wave allows you to manipulate audio files in a wide variety of ways and convert files into many different formats.

from an existing file or from one that you record using a microphone. The encoder includes controls that allow you to set the sampling rate, adjust the recording volume, add special effects such as echo, and edit the audio file using the cut, copy, and paste tools. Some encoders, such as the one that comes with Windows, offer a narrow range of controls. Others, such as Chris S. Craig's shareware program GoldWave (Fig. 5.6), are more like full-blown sound studios than simple recorders, allowing you to manipulate the audio file in virtually any way imaginable and save it in a variety of formats. You can even turn audio files recorded as mono and save them as stereo.

The *player* decompresses the audio file to its original size and plays it back. Like the encoder, the player offers its own set of controls. These controls allow you to do such things as start/stop or pause the recording, adjust the volume, and specify audio quality. In the case of Voxware's ToolVox Player (Fig. 5.7), you can even play back at a slower or faster rate without changing the pitch of the voice. This lets you slow down playback to improve comprehension or speed up playback to shorten listening time—with no distortion. As with other players, you can also move to a specific area of the audio file by dragging the player's slider bar.

The RealAudio Player not only lets you play back audio files you have recorded, but keeps track of audio files by title, author, and copyright (Fig.

Figure 5.7
Voxware's ToolVox
Player allows you to
play back files faster
or slower without
distortion.

5.8). This information is entered at the time files are recorded with the RealAudio Encoder.

Players have two roles. They enable you to review the audio file before sending it out as voice mail and they can be integrated into your Web browser as a helper application. When the browser encounters an audio file embedded in a Web page and it is in a recognized format, the player is loaded automatically and the file is played back. Once loaded, the player functions as a stand-alone program. (A more detailed discussion of so-called *streaming*, or real-time, audio and video applications is provided in Chap. 8, "Broadcasting Over the Internet.")

All audio data is defined by the following parameters:

- *Sampling rate.* In kilohertz (kHz) per second, typically ranging from 8 to 44 kHz. The higher the sampling rate, the greater the fidelity.

- *Number of bits per sample.* Typically 8 or 16 bits. Most sound cards and computers are equipped to handle 16-bit audio. While 8-bit samples are acceptable for recording voice mail, 16-bit samples are required for music and other sounds with rich content.

- *Number of channels.* Audio files are either monophonic (one channel) or stereophonic (two channels).

Figure 5.8
Progressive Net-
works' RealAudio
Player plays back
voice mail messages
and other RA-
formatted files and
helps you keep them
organized by title,
author, and copy-
right.

Most encoding programs allow you to set these parameters so that you can control both the quality of the recording and the amount of computer resources used. A stereo recording with a sampling rate of 44 kHz at 16 bits per sample is far more processing-intensive to create than a mono recording with a sampling rate of 8 kHz at 8 bits per sample. The latter is also smaller in size and is transmitted more quickly over the Internet.

You can save voice recordings in various formats. Some encoders even come with an editor or utility that allows you to convert the files into an appropriate format. This is useful if you get a lot of voice mail in different audio formats.

Encoders differ in the number and type of file formats supported. The most common audio file formats are:

AU. Audio, a Sun Microsystems format for 8-bit, monophonic (mono) sound

AIFF. Audio Interchange File Format, a Macintosh format for 8- or 16-bit, mono or stereophonic (stereo) sound

RA. RealAudio, a Progressive Networks format for 8- or 16-bit, mono sound

SND. System 7 Sound, a Macintosh format for 8- or 16-bit, mono or stereo sound

VOX. Voxware, a format for 8- or 16-bit, mono sound

WAV. Windows Audio format for 8- or 16-bit, mono or stereo sound

Because audio files are quite large, they must be compressed for efficient storage and transmission. Compression is achieved in software via a coder/decoder (*codec*). There are several compression formats in use today, some of which were originally designed for transmitting digitized voice over various type of networks, but which have been adapted for use on personal computers.

ADPCM. Adaptive Differential Pulse Code Modulation is the international audio communications standard recommended by the International Telecommunications Union (ITU) for voice over ordinary telephone lines. It offers 4:1 compression.

GSM. Global System for Mobile (GSM) telecommunications is an international compression standard for high-quality voice over digital cellular networks.

G.723. This is the international audio communications standard recommended by the International Telecommunications Union (ITU) for videoconferencing, PC telephony, and on-demand audio/video applications on the Internet and over regular telephone lines. It is based on DSP Group's TrueSpeech compression technology, which offers near toll quality voice.

TrueSpeech G.723 is based on an advanced algorithm that results in excellent voice quality despite the high compression rate. It operates at 6.3 and 5.3 kilobits per second (Kbps) with compression ratios of 20:1 and 24:1, respectively. TrueSpeech G.723 also includes a silence compression feature, which compresses out pauses between spoken words. Silence compression can bring the effective rate down to less than 3.7 Kbps.

There are also proprietary compression technologies such as Voxware's RT24 codec which is used in its ToolVox Encoder. The RT24 codec compresses speech files from WAV format to an 8 kHz, 2400 bps VOX file. The company's MetaVoice technology makes it possible to manipulate various elements of the human voice, including resonance, pitch, timbre, timing, and character, without distortion and without requiring large amounts of bandwidth, storage space, or processing.

MetaVoice works by creating two separate models of different vocal aspects: one creates a template of the user's vocal characteristics (spectral shape), the other records the user's vocal articulation and frequency. Spectral shape represents a *vocal map*, or *voiceprint* (the vocal characteristics that enable people to recognize the voices of others). Articulation and frequency are measurable, objective vocal elements, such as pitch and volume.

By altering one or both of these elements, MetaVoice is able to transform the human voice. For example, MetaVoice accomplishes *pitch shifting* by resynthesizing the voice at a new frequency, while leaving all other data intact. This makes it possible to shift a bass voice into soprano, for example, without distorting vocal character.

MetaVoice also allows users to optimize the three key elements of voice compression: bandwidth or file size, reproduction quality, and processing overhead. Three separate algorithms are available for different needs. Applications can use one algorithm or dynamically select from all three.

Current compression ratios range from 20:1 to more than 50:1. For real-time applications, quality is comparable to GSM-based algorithms that require two to four times the bandwidth. For 11 and 22 kHz playback, quality is comparable to 4:1 ADPCM compression. This performance can be achieved on the host PC processor without requiring specialized digital signal processing (DSP) support for compression or decompression/playback.

Sampling rate (kHz)	Audio frequency range (Hz)	Compression ratio	Data rate (bps)	File size bytes/hour of stored speech (MB)
8	50—4,000	53:1	2,400	1
11	50—5,500	20:1—35:1	5,000—8,500	3
22	50—11,000	35:1—55:1	6,500—10,000	4

Source: Voxware.

Voice Mail

Some e-mail programs offer the means to record and send audio files as attachments. In addition to Graphic E-Mail for Windows, there is Voice Mail from Bonzi Software and Pronto Mail from CommTouch Software. We can expect many more e-mail programs to support voice mail in the near future.

Pronto Mail, for example, offers an integral voice recorder that enables you to create a WAV file. Within the Compose window, you just click on the microphone toolbar icon to activate the built-in recorder and speak into your computer's microphone (Figs. 5.9 and 5.10).

Figure 5.9
Pronto Mail includes an integral recorder for creating voice mail messages. At the end of the recording, an icon appears at the bottom of the Compose window. By clicking on that icon, you can listen to the recording before sending it. If you are satisfied with the recording, just click the Send button.

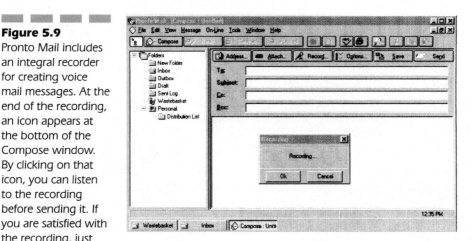

Figure 5.10

Pronto Mail puts an icon at the bottom of the Compose window indicating an attachment to the e-mail message.

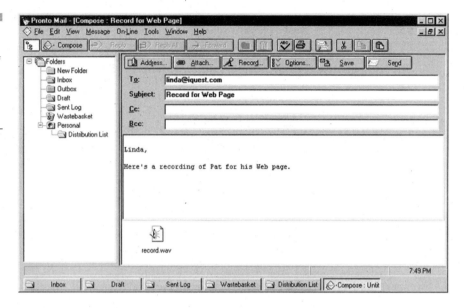

At the other end, another Pronto user opens the message and sees the same WAV file icon at the bottom of the message. Clicking on that icon plays back the recording. If the addressee does not use Pronto, the recording can be saved to disk and opened using an external audio player, such as the Media Player that comes with Windows.

Bonzi Software offers plug-ins that allow users to incorporate an audio recording and image of the sender into an e-mail message (Fig. 5.11). Plug-ins for America Online, CompuServe, Eudora, Microsoft Mail, and Netscape Mail are available for download from the company's Web site at

http://www.bonzi.com

VocalTec's Internet Voice Mail lets you record messages and send them to any Macintosh or Windows user with a valid e-mail address, including users who do not have Internet Voice Mail. You create and send voice mail messages using the following steps:

1. Type the recipient's e-mail address in the Address Panel.

2. Enter a subject (optional).

3. Record your message in the Voice Panel.

4. Add a text message (optional) in the Text Panel.

5. Add a file attachment (optional) in the File Attachment Panel.

6. Review the recorded message (optional).

Figure 5.11
Bonzi Software's Voice Mail plug-ins allow users of the most popular e-mail software programs and Web browsers to add an audio recording and image of themselves to their e-mail messages.

7. Click the Send button to send your voice mail message with text and file attachments.

Internet Voice Mail is available in Windows and Macintosh versions. You can download an evaluation copy of the software—good for 10 voice mail messages—from VocalTec's Web site at

http://www.vocaltec.com

A voice mail player comes with the software. You can send the player as an attachment (Fig. 5.12), so any recipient can hear your message. After a quick installation process, the receiver can listen to your message and save the player for future voice mail.

Recording Tips

To record the best quality voice mail messages, there are several considerations that merit close attention.

■ Minimize the amount of background noise, such as that coming from air conditioners, street traffic, fans, music, and talking.

■ Position your microphone about 4 to 5 inches away and speak slightly off center from the microphone.

■ If your microphone doesn't already have a wind screen, you can make one out of tissues or nylons and place it between you and the microphone to reduce pops from certain consonants (i.e., p, b, t, and so on).

Figure 5.12

With VocalTec's Internet Voice Mail program, you can send text messages and files in addition to recorded messages. Note the icon in the File Attachment Panel, indicating the presence of the Voice Mail Player program that will be sent along with the message. This allows the recipient to play the message, even if he or she does not have a copy of Internet Voice Mail.

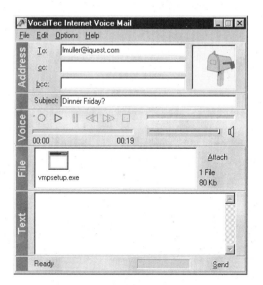

- If you are using an encoder which is optimized for the human voice rather than music—such as Voxware's ToolVox—record one person speaking at a time. If you plan to make use of sound effects and music in your voice mail, consider using another encoder, since compressing music, sound effects, and multiple voices with ToolVox will not produce satisfactory results.

- Use the highest sampling rate/frequency available. A WAV file with 16 bits per sample will sound better than an 8-bits-per-sample file after it is compressed.

- If you want to add special effects or music to your voice mail messages, try to avoid transferring sound files through multiple storage media (i.e., VHS to cassette to computer), since each transfer degrades the quality of the sound file.

Online Resources

The following table provides the Web links of the major vendors offering audio recorders and players from which you can download working or evaluation copies of their software. The Web pages also contain such information as platforms supported, system requirements, product fea-

Developer	Product	Web page or FTP site
Bonzi Software	Voice Mail	http://www.bonzi.com
CommTouch Software	Pronto Mail	http://www.commtouch.com
Chris S. Craig	GoldWave	http://web.cs.mun.ca/~chris3/goldwave/
DSP Group	TrueSpeech	http://www.truespeech.com
Graphic E Software	Graphic E-Mail for Windows	http://www.clever.net/ge/index.html
IRIS Systems	IRIS Phone	http://www.front.net/irisphone/iris.htm
Microsoft Corp.	Sound Recorder Media Player	http://www.microsoft.com
NetSpeak Corp.	WebPhone	http://www.netspeak.com
Progressive Networks	RealAudio	http://www.realaudio.com
VocalTec	Internet Phone (ver 4.0) Internet Voice Mail	http://www.vocaltec.com
Voxware	ToolVox	http://www.voxware.com

NOTE: This information, as well as updates, can be found at http://www.ddx.com/mgh.shtml.

tures, installation instructions, and troubleshooting advice. Since the technology is moving rapidly, you may want to consult these Web pages periodically for the latest developments.

Video Mail

Video mail is different from videoconferencing. Video mail uses the same store-and-forward method of delivery as e-mail, whereas videoconferencing is a real-time application that allows two or more people to see each other as they talk and share information. Both use similar technology for capturing images—a video camera mounted on or near your desktop computer and a microphone connected to your sound card. Video cameras connect either to your computer's parallel or serial port or to a video capture card. Connectix Corporation's QuickCam is an example of the former and Alaris's QuickVideo is an example of the latter.

Video mail can be used for personal messages or broadcasts or for short commercials about your company or product—all delivered over the Internet via e-mail. These messages can even be embedded in your Web pages with hypertext links. Upon activating the link, users can download the video message to disk for later viewing, or it can be opened instantly with a plug-in application registered with the Web browser.

Like voice mail, video mail entails the use of two software components: the encoder and the player. The encoder compresses video input so it can be transmitted over the Internet or stored on disk more efficiently. According to Alaris, one of its minute-long videograms is so highly compressed that it can be stored on a single floppy disk or posted on a Web page for instant playback, in addition to being sent over the Internet as an e-mail attachment. Alaris's Videogram technology uses a proprietary video compression algorithm to provide significantly higher compression ratios than existing compression algorithms including Indeo, Cinepak, and MPEG.* Combined with Alaris's speech compression algorithm TELP and music compression algorithm MUZIP, Videograms are able to achieve very small file sizes without significant loss in quality or frame rate.

The player can be any software that supports the format of the video files you create. If you record a video message in the AVI format, for example, a recipient can use any player that supports AVI to view the recording. Likewise, if the recording is compressed with MPEG, the recipient of your video mail must have an MPEG player to open it.

Since video mail software products use proprietary compression technologies, they usually come with a player. In the case of Alaris, a free Videogram viewer is available for download at its Web site:

http://www.alaris.com

Since the Videogram viewer can be embedded in the video message itself, you can distribute them to nearly anyone with a PC and Internet connection. The decoder simply self-extracts and self-executes whenever the video mail message is opened, so the recipient does not need any special hardware or software. And because the decoder is attached to the beginning of the message, users can cancel the downloading of a big file at any time and still view the video meesage up to the point where downloading stopped. Such self-playing files assure that the video messages will be viewed by the widest possible audience.

On the other hand, you do not really need special video mail software to send video mail. All you need is a digital camera to create the video, a compatible player to review the results, and an e-mail program that supports MIME attachments to send the file. Virtually any combination of

---

* MPEG (Moving Pictures Experts Group) is a compression standard for multimedia video applications. It was developed as an extension to the JPEG (Joint Photographic Experts Group) standard. Both are based on a mathematical technique called Discrete Cosine Transform (DCT) for representing sequences of data.

products will do. The recipient need only have a player that supports the same video format you made the recording in. However, voice mail software of the kind offered by Alaris and Connectix makes this a much smoother process, allowing you to create, review, and send video mail using a single program. And if you already have videoconferencing software, chances are it supports video mail as well.

MRA Associates, for example, offers a shareware version of VidCall, which not only provides videoconferencing and document sharing, but video mail. The free software, which can be downloaded from the company's Web page, works with about 25 types of video capture cards and accepts video from standard video cameras and VCRs and PC-oriented cameras such as the Connectix QuickCam.

The VidCall software is available for download at:

http://www.access.digex.net/~vidcall/vidcall.html

There are also card-based products that enable you to do video mail, videoconferencing, Internet video production, video editing and authoring, and CD-ROM title creation, all from a single add-in card. Array Microsystems' Personal Video Communications is an example of this type of solution. Array's offering enables the creation of video mail messages based on MPEG frame encoding. Because the technology incorporates store-and-forward capability, the existing Internet communications infrastructure can be used to send and receive video messages. For message recipients, the only requirements are an e-mail address and an MPEG-1 player, which comes standard on today's multimedia PCs. With scalable bit rate control from 100 Kbps to 2.4 Mbps, you can choose from among a range of quality and bit rate trade-offs to ensure efficient transmission and reasonable video quality.

System Requirements

Because the creation of video mail requires high-ratio file compression, the minimum resource requirement is a 486/33 computer with 4 MB of RAM. However, a Pentium computer with 16 MB of RAM is preferred because file encoding, which is usually more processing-intensive than decoding, will go much faster and will make for less frustration during file creation. For example, with some video products compression is done entirely in software with no help from a video capture card. In this case, it can take from two to six times the actual playing time of the file to completely encode it. A good rule of thumb here is, if you use a video capture card that does the encoding, you can get away with a 486 computer. But if you use a software-only encoding product, you will be better off with a Pentium computer.

There are numerous types of digital cameras available to choose from with a price range of $100 to $1000 or more. The gray-scale QuickCam camera from Connectix can be purchased for less than $100 from various mail-order companies. Instead of requiring a video capture card, the camera simply plugs into the computer's parallel port. It takes still shots and records video at 10 frames per second, which means the picture is always jerky and your voice tends to run way ahead of your lips and head movements.

A color version of the popular cue-ball sized camera offers much better picture quality at 24 frames per second, but is double the price. By contrast, full-motion video without these distortions requires 30 frames per second, but this is usually achieved with the aid of MPEG-based graphics accelerator cards. The color QuickCam includes such features as a manually adjustable focus lens, a self timer for taking photographs, and automatic video capture for live Web images. It also automatically adjusts for brightness and hue. Unlike other PC-based cameras, however, QuickCam does not include an integral microphone.

The FlexCam Pro from VideoLabs costs about $350 from mail order. This color camera is mounted on a flexible 18-inch gooseneck stand and includes a built-in microphone that can record in stereo. The long stem is very handy, since any video camera usually must be manipulated to capture just the right amount of light for a good quality image. The FlexCam plugs into a video capture card such as the company's $200 Stinger or $400 QuickVideo.

StarDot Technologies' WinCam One, which sells for about $300, is a gray-scale camera that takes still shots as well as video. The camera plugs into a PC's serial port and can be located up to 250 feet away from the PC. This allows you to be more mobile, taking the video in a different room, possibly with a group of people, instead of being bound to your desk.

If you really want to get serious with video mail, there are even higher priced cameras to choose from. Sony Electronics, for example, offers the PC CAM, a desktop video camera and microphone intended for multimedia program creation, which is priced at about $500. It features an 11-inch neck and an adjustable camera iris to compensate for lighting conditions, but the microphone is monophonic.

Ricoh offers a versatile digital camera that can record full-motion video (but only 5 seconds' worth) and audio, audio alone, and other combinations, including stills with audio, rapid-fire stills, and even a special mode for taking close-ups of documents. The base price for the RDC-1 is about $1600, but a full-featured configuration can cost well over $2000.

Some monitors are available that are specifically designed for multimedia. Nokia's Multigraph 17-inch 447 series of monitors, for example, include a built-in video camera, microphone, and a pair of speakers for about $1500. Other models are available with smaller or larger screens.

Companies such as IBM, Zenith Data Systems, and Unisys sell desktop computers that are completely outfitted for videoconferencing and, consequently, video mail. They come with a camera, microphone, and videoconferencing software—all of which have been pretested for compatibility to save you the hassle of trying to build a system from scratch. In addition, most major vendors of notebook computers have multimedia models that come with CD-ROM, built-in speakers, and microphone. All you need for video mail (or videoconferencing) is to attach a device such as the QuickCam to the parallel port.

Be aware that the cost of a camera and/or video capture card is not always a good indicator of video quality. Since the technology is moving by leaps and bounds, you should check the reviews of the major PC and Macintosh magazines before spending any serious money. These reviews usually include test results and comparisons of vendor offerings. If you are unsure of how frequently you will use video mail and just want to dabble with the technology, Connectix's QuickCam is the most economical and easiest to set up and use. You can graduate to something more sophisticated later as your interests or needs change.

Some vendors offer product bundles that include everything you need to create and edit video mail. VideoLabs, for example, offers a complete multimedia package that includes a high-resolution, full-motion video camera, Windows-based video capture board/frame grabber, and image-editing software.

You don't need a special video camera for your desktop computer; you can use your existing camcorder/VCR to create and view high-quality video clips. Since your camcorder is not tethered to your computer, you have the added advantage of unlimited mobility. You can shoot in any room or even outdoors. In addition, you have available a number of features that enable you to control video quality while you are shooting. Some camcorders even allow you to play back your clip immediately after shooting it. This gives you a chance to do a quality check and shoot the video again if necessary. These features can be a real time saver, especially when all you're trying to do is prepare a short clip to send as video mail.

After shooting your video, you can use a program such as Gold Disk's VideoDirector to edit videos from camcorder to VCR. You control the units from the computer with a cable connected to the serial port. When you are pleased with the results, you can bring the video to your hard disk using any video capture card that supports NTSC and PAL* and your video recorder's format (i.e., VHS, S-VHS, Video8, and Hi8). Once the video clip is captured in RGB 24-, 16-, or 8-bit color, it can be saved in the Windows AVI format before sending it as an e-mail attachment. VideoDirector lists for only $60.

A more expensive solution is the desktop video capture and editing system offered by TV One Multimedia Solutions. Its MovieX system is a complete package for digital video recording, playback, and editing. It consists of a JPEG card with video capture software, digital editing software, and device control software. You can capture video up to 640 × 480 resolution at 30 frames per second in NTSC and 25 frames per second in PAL. The ISA version of MovieX is capable of a minimum compression ratio of 9:1 and approaches S-VHS videotape quality. The PCI version is capable of a minimum compression ratio of 5:1 and exceeds S-VHS videotape quality. MovieX is a more sophisticated product and would be used for more than just video mail. Prices range from $895 for the ISA version to $1495 for the PCI version.

There are about a dozen video capture and editing systems for Windows and Macintosh computers that can be adapted for use in preparing video mail. The number of vendors is growing. You can find the Web pages of these companies by entering the term *video capture* in your favorite Internet search engine.

Sending Video Mail

To demonstrate the ease with which video mail can be created and sent, this section discusses the use of Connectix VideoMail, which is a capability of the company's VideoPhone software package for videoconferencing.

When you open the VideoMail application, you will be presented with the Preview window (Fig. 5.13). To record a video message, you can simply press the space bar or select the Movie menu, then choose Record.

Next, the Capture dialog box will appear, displaying the current video capture settings (Fig. 5.14). To start recording, you click on the Begin button. To stop recording, you click on the End button.

When you have finished recording, the video is stored in temporary workspace until you save it to disk. Although you do not need to save

---

* NTSC (National Television Standards Committee) and PAL (phase alternative line) are television picture standards for North America and Europe, respectively.

■■ ■■ ■■ ■■

Figure 5.13
The Preview window in Connectix Video-Mail.

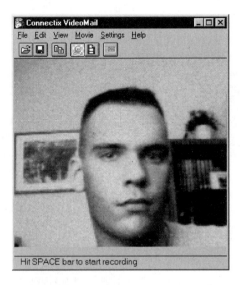

your video to send it, you have the option of saving it for transmission at a later time.

After creating a video message, you can review it by clicking on the Playback button. If you are not happy with the results—perhaps the audio or video settings were not quite right—you can discard it by pressing the space bar. A dialog box will appear, asking if you want to save your work. Clicking No discards the video message and brings up the Capture dialog box again so you can record a new message.

If you had a problem with poor audio, you can adjust it by choosing Audio from the Settings menu, then choosing Audio Format. You will be able to select one of three quality settings:

■ *CD-quality sound.* Provides the highest-quality sound

■ *Radio-quality sound.* Provides medium-quality sound

■ *Telephone-quality sound.* Provides telephone-quality sound

If you had a problem with poor video, you can adjust it by choosing Video from the Settings menu. You can adjust the size and quality of the

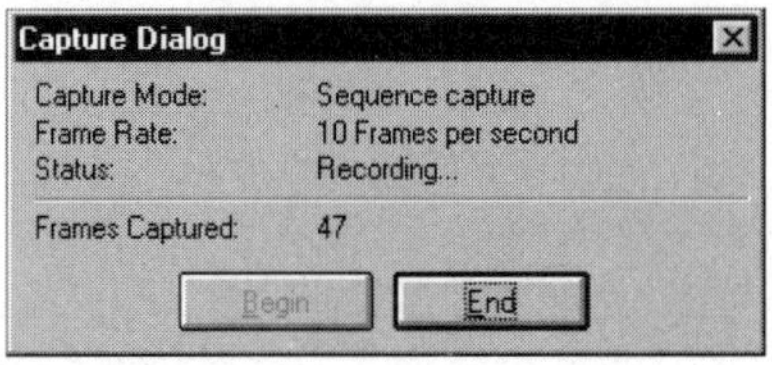

■■ ■■ ■■ ■■

Figure 5.14
The Capture dialog box in Connectix VideoMail.

video image, adjust the brightness and contrast, and adjust the rate at which frames will be captured. There are three default settings for frame capture:

- 5 frames per second
- 10 frames per second
- 15 frames per second

You can also specify another frame rate by selecting Other and entering a number in the dialog that appears. The higher the frame capture rate selected, the better the video quality but the larger the file will be and the longer it will take to send over the Internet and the longer it will take for your recipient to download. These factors should be taken into consideration when recording video messages, so you can select only the audio and video quality levels you really need.

After the adjustments are made and you have recorded a new video message, you can review it by again clicking on the Playback button. Assuming you are now happy with the results, it is time to send your video message. If you are using the VideoMail feature for the first time, you must configure the mail settings. This is done by selecting Mail from the Settings menu, which brings up the Mail Configuration dialog box (Fig. 5.15) where you enter your name, e-mail address, and SMTP server.

To send your video message, click the Send As E-Mail button while in the Preview window or Playback window. In the Send dialog box that appears, enter an address (or select an address from your list), and click Send to transmit the video message. It's that simple.

If you're wondering if video mail can be sent over such online services as America Online and CompuServe, the answer is yes. You simply create the video mail as described above and send it to anyone on any service as a message attachment.

Figure 5.15
The Mail Configuration dialog box in Connectix VideoMail.

Mail Configuration   ☒

Please enter the following mail information:

For the Return Address, enter YOUR Email address.

For the SMTP server, enter your SMTP server name,
such as mail.company.com, or the servers IP
address (within brackets), such as
[123.123.123.123].

Your Name: | Matt Muller |

Return Address: | mmuller@ddx.com |

SMTP Server: | mail.iquest.com |

      [OK]   [Cancel]

An interesting offering from Connectix is QuickCards, which is a multimedia greeting card package. It includes a CD-ROM with over 60 predefined templates (i.e., cards), which you can personalize by adding your own pictures, movies, and text to the rich content of each card. A finished QuickCard can be e-mailed for playback on a Macintosh or PC or simply saved to a floppy disk and mailed to the recipient.

Production Tips

There are a few simple things you should pay attention to that will improve the effectiveness of your video mail messages:

- *Lighting.* You should have a light source in front of you to illuminate your face and eliminate hard edges, rather than have the light source above or behind you.

- *Background.* Pay attention to your background. If it is cluttered and messy, you could be conveying a poor image. If you record a lot of video messages, consider rearranging your room or making a portable screen to put in back of you. A screen has the added benefit of blocking the movements of others, which may cause distraction.

- *File size.* Make sure the audio and video settings are adequate to ensure a good quality video message, but not such that file size will be unnecessarily inflated. Don't forget, the recipient will have to download the file before opening it. If the file is too big, the recipient may cancel the download and your message may go unseen.

- *The message.* You should know what you want to say before you say it so the viewer's time will not be wasted. Outline what you want to say and hang it on the wall in back of your computer. You can refer to the outline while looking as though you are talking into the camera—sort of a pauper's teleprompter.

- *Shooting angle.* With PC-mounted cameras, mug (straight on) shots are unavoidable, but they are also boring. So consider sitting at an angle to add perspective or hold a prop to add interest.

- *Feature usage.* If you use a camcorder to record video mail messages, be careful not to overuse such features as pan and zoom. While these features can be used to add interest, overuse can detract from your message and focus the viewer's attention instead on the camera's movement. Furthermore, to come off well, these features require a higher frame rate, which increases file size.

▧ *Review before sending.* It is always a good idea to review the video message before sending it out. This not only lets you check audio and video quality, but the screen for unwanted distractions that may have happened while recording, such as the prankster in the background who gets a kick out of waving at the camera incessantly or gesturing a profanity at every opportunity.

If you are preparing video messages for coworkers, clients, or potential customers you will want to project a professional image. Getting better equipment can help improve audio and video quality, which can translate into projecting a better business image. If you can't afford better equipment, following these simple guidelines can go a long way toward making up for that.

Online Resources

The following table provides the Web links of the major vendors offering video mail solutions from which you can download working or evaluation copies of their software, if available. The Web pages also contain such information as platforms supported, system requirements, product features, installation instructions, and troubleshooting advice. Since the technology is developing quickly, you may want to check these Web pages periodically for the latest information.

Developer	Product	Web page or FTP site
Alaris	Videogram, QuickVideo	http://www.alaris.com
Array Microsystems	Personal Video Communications	http://www.array.com
Connectix	QuickCam, VideoPhone	http://www.connectix.com
MRA Associates	VidCall	http://www.access.digex.net/~vidcall/vidcall.html
Silicon Graphics	MediaMail	http://www.sgi.com/
TV One Multimedia Solutions	MovieX	http://www.tvone.com
Darald Trinka	VideoMail Pro	http://www.shout.net/~dtrinka/home.html

NOTE: This information, as well as updates, can be found at http://www.ddx.com/mgh.shtml.

Conclusion

Audio and video can be much more subtle than text and graphics in conveying your message and can do it with more flair and impact. Research has shown that audio and video can significantly increase the retention of information among listeners and viewers, making this kind of messaging especially useful for business purposes. However, the same benefits can be achieved using voice and video mail for personal communications with family and friends, who will always look forward to hearing and seeing you—particularly when you cannot always be with them on important occasions. In such cases, the Internet can be the next best thing to being there!

Faxing Over the Internet

Introduction

In the United States, faxing accounts for 30 percent of corporate telecommunications bills, according to an annual study conducted by The Gallup Organization and Pitney Bowes Facsimile Systems Division. Fortune 500 companies alone spend an average of $15 million annually on fax-related transmission charges, with expectations that this number will increase by 12 percent a year. In other countries, faxing accounts for 40 to 65 percent of the bill due to higher telecommunications costs. Consequently, there is every incentive for large companies, home-based businesses, and individuals to leverage their existing Internet connections by using them to send faxes just as they do to send e-mail.

The reason for the high cost of faxing is that each document must be scanned into an image, which greatly inflates the number of bytes that must be sent. Even with compression, it takes a long time to transfer the image through the network. Since charges are based on the duration and distance of the call, the cost of faxing adds up fast. It can cost several dollars to send a single page from New York to Tokyo, for example, during normal business hours, depending on the speed of your modem/fax and the quality of the connection.

With such costs, you might wonder why e-mail isn't used instead. After all, it is virtually free over the Internet and new services are becoming available that offer e-mail at absolutely no cost and with no need for a connection to the Internet (see Chap. 4 for details). However, many documents are not in electronic form. In fact, they may contain annotations, signatures, drawings, or clippings from other sources that are important to retain in their original form, even if they can easily be put into electronic form for transmission as e-mail. Other times, you may want to send messages and documents economically over the Internet to people who do not have access to the Internet, e-mail software, or even a computer. All they might have is a stand-alone fax machine. Thus, there continues to be heavy reliance on facsimile, not only among large companies, but telecommuters, home-based businesses, and individuals as well.

The high cost of faxing, its inordinate share of the total telecommunications bill, and its continuing necessity provide ample justification for looking to Internet-based solutions that offer opportunities for cost containment.

There are several ways to send faxes over the Internet. You can choose from among several low-cost mail-to-fax gateways offered on a subscription basis, use a technique called remote printing, or buy very inexpensive Windows 95 application software. Each of these methods has advantages and disadvantages. The choice will depend on several factors, including geographical reach, cost per fax (if any), the number of faxes you send a month, and ease of use.

Mail-to-Fax Gateways

Mail-to-fax gateway services are springing up everywhere. Basically, they give you the ability to send faxes as you would e-mail. This is an important capability to have, considering that there are more than 200 million fax users worldwide who do not use e-mail. The software needed to send faxes in this manner is usually available free and you can try out the service before you commit to a subscription. Among other things, these services differ by features, pricing, and geographical coverage.

FaxSav

One of the companies that provides fax services over the Internet is FaxSav, an operator of a dedicated worldwide fax network that provides discounts over phone lines. The company also offers a service that allows users to access its fax network via the Internet. This service is of particular value to users in other countries because it allows them to bypass the often costly PTT networks to send faxes to the United States using their existing Internet connections. The service can save users in other countries as much as 80 percent on faxes to the United States.

FaxSav offers two free software products that are used for faxing over the Internet: FaxLauncher, which gives Windows users the ability to fax directly from an application, and FaxMailer, which allows users to convert electronic mail to faxes to reach conventional fax machines that are not connected to the Internet. For a limited time, both software products are available for free use and can be downloaded from the FaxSav Web site at:

http://www.faxsav.com

The cost of sending faxes using these applications is a fixed $0.15 per page for delivery anywhere in the United States. (The same rate applies to faxes sent within any other country.) Examples of basic rates per page for delivery from the United States to other countries are as follows: Australia $0.39, Germany $0.49, Japan $0.45, and the United Kingdom $0.35. As FaxSav adds more local delivery nodes around the world, it expects to reduce prices even further to $0.20 to $0.25 per page.

Internet delays are minimized by the architecture of the FaxSav network (Fig. 6.1). After the short, initial transmission from your desktop to the nearest FaxSav server on the Internet, FaxSav takes over the fax delivery process. FaxSav uses multiple types of networks as well as the Internet to dynamically select the best overall route for delivering each fax. The use of multiple networks gives FaxSav worldwide coverage, so you don't have to worry that your destination area code isn't on the network.

As documents traverse the Internet, confidentiality is assured by encryption supplied by RSA Data Security. Confirmation of delivery is made via e-mail. If delivery cannot be completed, an error message is returned to the sender via e-mail explaining the reason.

FAXLAUNCHER. FaxLauncher is used to send a fully formatted fax over the Internet from any Windows application, provided that your computer has an Internet connection. FaxLauncher can be added to any Windows application by using the procedure to add a printer driver (Fig. 6.2).

After you register with FaxSav for trial usage, you can simply go to your favorite application and open the document you want to fax. Then select Print from the File menu and choose the FaxLauncher driver. When you click OK, the FaxSav Delivery Information window will pop up (Fig. 6.3). This is where you enter the fax number (include country code, area or city code, plus the fax number), subject, and contact information. The information entered here will be used to compose the default cover sheet.

The document you opened in the first place will be transmitted after the cover sheet upon clicking OK. At this point, the FaxLauncher Delivery window pops up, showing you the transmission status of the document (Fig. 6.4).

At the receiving fax machine, the document arrives just as it would from any other fax machine (Fig. 6.5). At the bottom of the cover page, there is an advertisement indicating that the fax was transmitted via FaxSav and showing the company's e-mail address and Web page.

When the fax is successfully delivered, FaxSav sends back a delivery notice via e-mail that looks something like this:

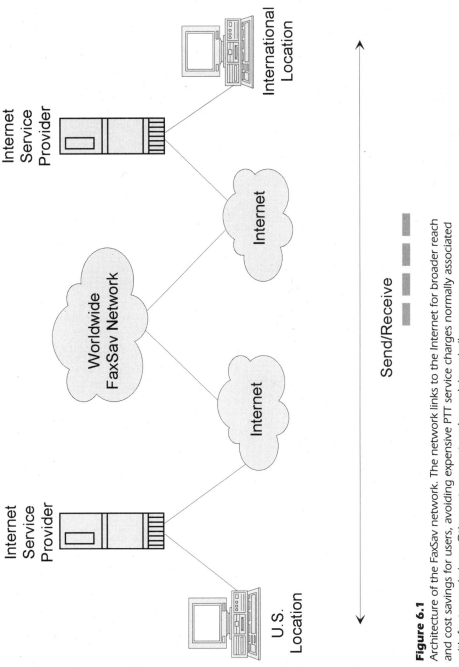

Figure 6.1

Architecture of the FaxSav network. The network links to the Internet for broader reach and cost savings for users, avoiding expensive PTT service charges normally associated with fax transmissions. Other gateway networks work in a similar manner.

Figure 6.2
In Windows 95, the
FaxLauncher software
will automatically
install a new printer
driver called
FaxLauncher. When
you select Print from
Word for Windows
7.0, for example,
FaxLauncher is one of
the printer drivers
you can select.

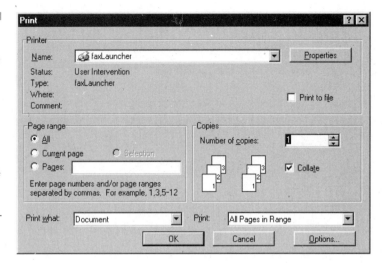

```
Date: Mon, 15 Jul 1996 21:03:13 -0400 (EDT)
From: support@faxsav.com (faxSAV)
Subject: FaxSav delivery notice: Thank you!
Apparently-To: <nmuller@ddx.com>


DELIVERY NOTICE


Your email or fax received at July 16, 01:01:59 AM GMT from email
address nmuller@ddx.com with a subject of <Thank you!> was
delivered to 12058807159.


The date and time of delivery was July 16, 01:03:09 AM GMT. The
number of pages delivered was 2.


Delivery was made on attempt # 1.


Called fax machine identifier:
Your Fax Document Reference #: 9197757330000


At our United States rate, this fax would cost $0.30 US, but due to
our free fax promotion, there is no charge. (Price quoted does not
consider volume discounts or other customized rate plans).


Thank you for using FaxSav for Internet faxing services. For
information on the FaxSav family of global communication solutions,
contact FaxSav Incorporated: by email at faxsav@faxsav.com on the
WWW at http://www.faxsav.com or by phone at: 1-800-828-7115
(within the U.S.), or +1 908 906 1555 (outside the U.S.)
```

Figure 6.3
FaxSav Delivery Information window. Clicking OK at this point launches the fax.

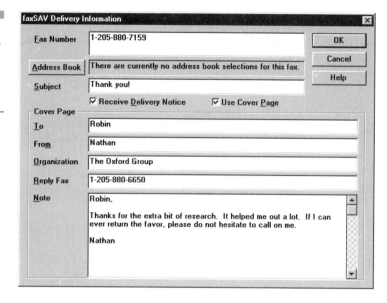

FAXMAILER. FaxMailer extends the capabilities of all e-mail packages, in essence giving every fax machine a virtual e-mail address. By merely changing an address field in any Internet-accessible e-mail package, you can send e-mail to any fax machine. This is particularly convenient for reaching people who do not have e-mail accounts. This is done by entering the fax number of the remote device in the To: field of the e-mail program using the following format:

faxnumber@faxsav.com

where faxnumber is the telephone number of the remote fax machine. You must also enter your unique FaxMailer stamp at the end of your

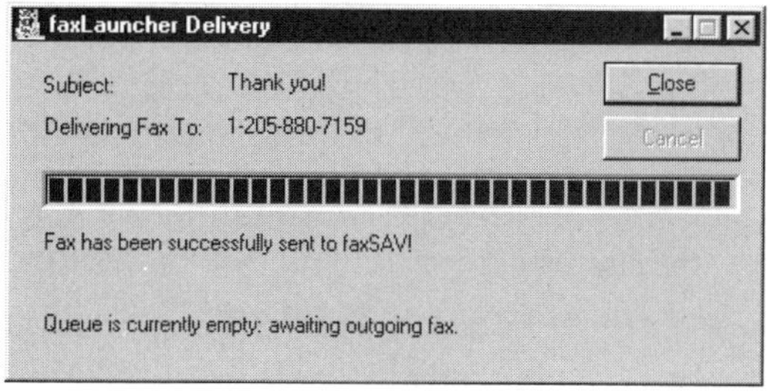

Figure 6.4
FaxLauncher shows the transmission status of the document. In this case, the fax has been successfully sent to FaxSav and the queue is empty, awaiting another outgoing fax.

Figure 6.5

The fax as delivered
via FaxSav and output
from a stand-alone
fax machine.

Internet Fax Delivery

To:	Robin
Fax Number:	12058807159
From:	Nathan
Organization:	The Oxford Group
Fax Number:	1-205-880-6650
Date:	July 16, 01:01:59 AM GMT
Subject:	Thank you!
Pages:	2 (including this cover sheet)

Note:

Robin,

Thanks for the extra bit of research. It helped me out a lot. If I can
ever return the favor, please do not hestitate to call on me.

Nathan

 This document was transmitted via faxSAV for Internet, another global communications solution from the faxSAV family of services. Contact faxSAV Incorporated at internet@faxsav.com or visit www.faxsav.com for more information on how you can save.

message on the last line by itself (Fig. 6.6). *Stamps* are a method of validating that e-mail and faxes come from you and not someone else claiming to be you. You then click on the Send icon and FaxSav delivers the fax.

The stamps are free and are good for 20 faxes. You are automatically sent a new stamp by e-mail when the previous stamp is almost used up. If you try to send a message using an expired or invalid stamp, FaxSav sends

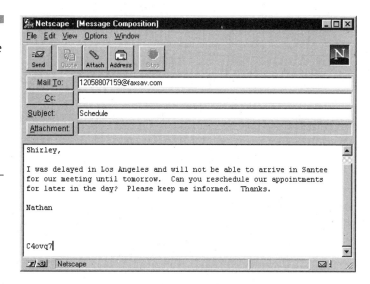

you a failure notice via e-mail informing you that the message was not delivered and for what reason.

FAXSCAN. FaxLauncher is compatible with Windows-based, scanner utility software and includes a Twain-compatible plug-in called FaxScan to conveniently send scanned documents through the Internet to the FaxSav network, in essence creating a virtual fax machine at the desktop. If your scanner utility can send a scanned file to print, then it can send a scanned file over the Internet to FaxSav.

FaxSav has many improvements planned. With regard to FaxLauncher, there will be a resend capability, in case the remote fax machine is busy; offline fax queuing, allowing you to assemble multiple faxes for transmission at a convenient time; a log of sent, delivered, and undelivered faxes; and the ability to import address book information from other applications. With regard to FaxMailer, there will be greater attachment support, especially for graphics. At this writing, FaxMailer supports Post-Script, HTML, TIFF, JPEG, and GIF files as attachments.

When you first register with FaxSav, you will receive 10 free faxes. Once your first 10 faxes have been used, the company bills your credit card a fee of $15, which allows you to fax up to 100 pages within the United States. When the maximum amount of pages have been faxed, the company bills your credit card another $15. Accounts that regularly fax more than $15 per month can set their billing fee at a higher rate to reduce the number of multiple charges to their card in a single month.

Faxaway

International Telcom's Faxaway is another mail-to-fax gateway. It allows users to send a fax anywhere in the United States from anywhere in the world for as little as $0.05. To use the service, you compose your fax in the form of an e-mail message, using whatever software you normally use for composing and sending e-mail (Fig. 6.7). No extra software is required.

Faxaway uses the same addressing scheme as FaxSav: faxnumber@faxaway.com which would include the country code and area code of the recipient. For example, the country code for the United States is 1; thus a fax to Faxaway in the Seattle area would be sent to

12065553434@faxaway.com

The e-mail arrives at the company's nearest Internet node where a telephone line is opened to the remote fax machine specified by the telephone number you entered. The e-mail is then received at that fax machine. You will receive a confirmation by e-mail from Faxaway once your fax is transmitted.

Faxaway charges for each fax in 6-second increments with a 30-second minimum. On average, approximately 10 lines of text can be sent in 30 seconds, depending on the speed of the receiving fax machine. You pay only for transmission time, based on the company's worldwide rates. Faxing to a fax machine in the United States could cost as little as $0.05. There are no setup or monthly charges. A prepayment plan is available as well

Figure 6.7
You can try out Faxaway's service at http://www.faxaway.com. The demonstration calls up Netscape Navigator's built-in e-mail program which can be used to send an e-mail message to any fax machine. You can send $5 worth of messages free.

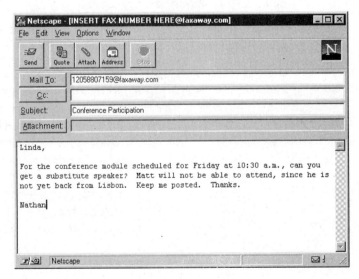

under which Faxaway charges your credit card up to a specified limit. For example, if you set up a monthly limit of $100, that means that in one month, you could send $100 worth of faxes and no more. With every message you send, you get back a confirmation receipt that tells you how much the transmission cost, plus how much credit you have remaining.

Faxaway supports sending attachments via UU-encoding (UUE), MIME, and BinHex. In addition, Faxaway supports the following attached document types: Microsoft Word, WordPerfect 5.x, Lotus Ami Pro, text files, bitmap files (BMP), and JPEG files. Attachments cannot be sent as part of the free fax offer.

You create your own fax cover page in one of the supported word processors, adding graphics, special fonts, fields, headers, a letterhead, and a signature. Faxaway then attaches this document to your account, so that all of your faxes have a similar look. There is a $10 installation charge per custom page.

Faxaway also supports broadcast faxing. You can compose, edit, and append an unlimited number of broadcast fax groups, containing individual names, companies, fax numbers, and other data. Users give each broadcast fax group a unique name. Users take a document and send it as an e-mail attachment to groupname@faxaway.com, where *name* refers to the unique name of the broadcast fax group. The document will merge with information from the broadcast fax group before Faxaway sends it to all of the fax numbers in the broadcast fax group.

A complete online guide to using Faxaway can be found at the company's Web site at:

http://www.faxaway.com/guide.html

Faxscape

Faxscape's mail-to-fax service allows you to send documents to anywhere in the world for a flat rate of $0.75 per page. Transmissions are handled by an Australian company, Newsnet ITN. You simply open an account, purchase fax pages, and send your faxes.

Aside from a flat rate per page to all countries, Faxscape offers the following key features:

- Immediate authorization and service
- Secure transmission over a global fax network
- Secure Credit Card Transfers using SSL (Secure Sockets Layer)

- E-mail confirmation of fax transmissions
- Resend capability that attempts to send faxes to busy and out-of-paper fax machines for up to 12 hours.

If Faxscape encounters a voice recording at the remote location, it will wait until the recording is finished before sending the fax. If the receiving fax machine is also a phone, Faxscape will automatically redial until the remote machine is switched to the fax mode.

Faxscape's WEB-FAX Service lets you fax from within Netscape Navigator, which takes advantage of SSL to ensure secure transmissions. Internet users can access the Faxscape secure WEB-FAX Service directly from the company's web site at:

http://www.faxscape.com

All successful Faxscape WEB-FAX transmissions are logged into a personal Web-page journal for your future reference. Confirmation of fax delivery is sent to you via e-mail. If there has been a delay because the remote fax machine is busy, out of paper, or not answering, you will be notified by e-mail of delivery progress.

At this writing, Faxscape had announced its intention to offer a free printer driver for Windows 3.1 and Windows 95. Called NET-FAX, the printer driver will allow faxing from within any Windows-based application.

JFAX Communications

This company's JFAX Personal Telecom service operates slightly differently from other gateway services. Instead of mail to fax, you have the option of subscribing to the company's fax-to-e-mail service. With this receive-only service, your contacts can fax you and leave voice mail messages on your unique personal JFAX telephone number. You can sign up for a number in any city on the company's global network. The network is still quite new, so the number of cities is limited. Internet delivery of faxes can be made to Atlanta, Chicago, Hong Kong, London, Los Angeles, New York, Paris, San Francisco, Sydney, Tokyo, and Washington. JFAX also offer users a U.S. toll-free service.

JFAX uses proprietary technology to instantly convert incoming faxes into a compressed graphic binary file and automatically delivers it via the Internet to the subscriber's e-mail box. When you check your electronic mailbox, the faxes are there in their original format with good graphic and text quality.

Your incoming faxes and voice mail are delivered to you via e-mail, no matter where you are. The sender only has to know your fax number. You don't have to be online when a fax or voice mail message comes in, since they will be waiting in your e-mail inbox the next time you log on. Since JFAX delivers faxes and voice messages directly into a subscriber's e-mail box, this eliminates having to forward faxes to offices or hotels where they can be read by others.

The service eliminates the need for a stand-alone fax machine (and associated supplies) to receive faxes as well as an answering machine or voice mail service. All you need is JFAX Communicator software to view your faxes and hear your voice mail. Faxes arrive as attached files in their original format. You can retrieve them with the free JFAX Communicator software for Windows 3.x, Windows 95, or Macintosh. The software interfaces with e-mail and online service packages and allows users to zoom in and out, rotate or adjust the size of the document to get a full view of incoming faxes. The voice-mail-to-e-mail capability is available at no extra charge to all subscribers. The voice mail component utilizes a compression algorithm to facilitate the transmission of recorded voice messages over the Internet.

The JFAX Communicator software is free and the service costs $12.50 a month for up to 100 pages of incoming faxes per month. Additional pages cost $0.20 each. There is a $30 account activation fee as well.

The company also offers a service that allows you to send faxes via e-mail. JFAX FaxSend allows you to send faxes via e-mail to any fax machine from within your favorite applications. JFAX users who sign up for the fax-to-e-mail service can use FaxSend without any sign-up fee and without any added monthly service charge. The only cost will be for the actual faxes you send, and at greatly discounted rates.

Internet users are able to sign up for these JFAX services and download the free software on the company's World Wide Web site at:

http://www.jfax.net

NetCentric's FaxStorm

NetCentric Corporation offers fax software called FaxStorm, which gives Internet users the means to send point-to-point and broadcast faxes to any fax machine worldwide for as little as $0.02 per minute. FaxStorm products can be downloaded from NetCentric's Web page at:

http://www.netcentric.com

Among the noteworthy features of FaxStorm is that the application provides real-time status on the progress of fax transmission. FaxStorm's status manager uses proprietary call progress technology, providing confirmation of delivery and automatically resending the fax if necessary. This eliminates repeated visits to the fax machine, waiting for a redial due to a busy signal, no answer, or lack of paper on a receiving machine. FaxStorm also ensures the confidentiality of fax contents by employing RSA Data Security's BSAFE product as the encryption engine to ensure security and authentication. (BSAFE includes modules for popular encryption techniques, such as RSA, DES, RC2, and RC4, and also supports digital signatures and certificates.)

NetCentric offers users of Windows 3.1, Windows 95, or Windows NT several versions of FaxStorm that work over the Internet:

- *FaxStorm Desktop.* A full-featured PC-based faxing application that includes a contact manager, fax scheduler, and cover page designer. In addition to point-to-point faxing, FaxStorm Desktop enables broadcast faxing.

- *FaxStorm SoftModem.* This software gives users of popular PC-based faxing software such as Delrina WinFax and Microsoft Exchange the means to send faxes over the Internet.

- *FaxStorm Web.* This is a Web-based faxing solution that uses browsers such as Netscape Navigator. This would be used by organizations or individuals who do not use Microsoft Windows.

- *FaxStorm Print Driver.* You can fax over the Internet directly from any Windows-based application by selecting the FaxStorm printer driver.

FaxStorm uses NetCentric's POPware software installed inside the Internet at the points of presence (POP) where the phone network and the Internet meet. This enables communications such as faxes to be delivered over the Internet to computer users and through phone lines to everyone else. FaxStorm compresses, encrypts, and transmits faxes to a POPware server. Once the fax has been launched, NetCentric's POPware servers use intelligent routing algorithms to send faxes over the Internet to the NetCentric POPware server closest to its final destination. It then enables the server to dial a call off the Internet to the destination fax machine.

If you happen to be an Internet service provider (ISP), NetCentric offers the means for you to offer fax services to your subscribers. The company's POPware runs on Pentium-based workstations installed at ISP points of presence, making it possible to reach out to local fax machines via dial-up links. Customers dial into a local ISP POPware server and upload a fax file that is routed to the POPware server nearest the target destination and then delivered via a dial-up link.

FaxStorm's printer driver works like other products such as FaxSav in that you simply select the driver within an application whenever you want to fax a document. When you are ready to send a fax, you fill out a pop-up Send Fax form (Fig. 6.8) with information that goes on the cover page (Fig. 6.9). If you have attachments, you select the Attachments tab to specify which document should follow the cover page.

FaxStorm's pop-up status manager shows you the status of faxes: completed, queued, and transmission progress (Fig. 6.10). You can sort faxes by creation time, view faxes already sent, and view status details.

After the fax is sent, FaxStorm provides a comprehensive status report about the transmission in a pop-up Fax Detail window (Fig. 6.11).

Another useful feature of FaxStorm is its Contact Manager (Fig. 6.12) which is essentially an address book containing the names, companies, fax numbers, and phone numbers of the people you contact most. A built-in search engine lets you find entries by any of these categories.

Remote Printing

The mail-to-fax gateway services described above are enhanced versions of the experiment in remote printing, which has been available to Internet

Figure 6.8
FaxStorm's Send Fax window allows you to specify cover page information, attachments, and options. Options include the abilities to add billing codes to faxes, set the resolution of faxes, schedule the delivery of faxes, and specify what sender information should show up on the top of each page sent.

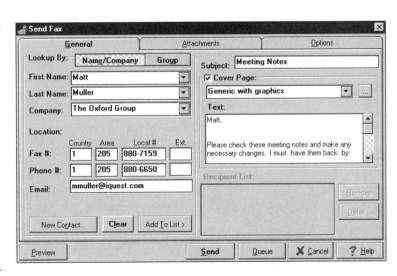

Figure 6.9
FaxStorm delivers the
fax via the Internet to
the remote machine.
The default cover
page includes all the
information specified
in the Send Fax win-
dow—and no adver-
tising shows up in
the pages output at
the remote machine.

TO:	**Matt Muller**
COMPANY:	**The Oxford Group**
PHONE NO.:	**1 (205) 880-6650**
FAX NO.:	**1 (205) 880-7159**
FROM:	**Nathan Muller**
COMPANY:	
PHONE NO.:	**1 (205) 880-6650**
FAX NO.:	
NO. OF PAGES:	**2**
SUBJECT:	**Meeting Notes**

Comments:

Matt,

Please check these meeting notes and make any necessary changes. I must have them back by Wednesday if we are to meet the filing deadline.

Thanks.

users on a limited basis since mid-1993.* The purpose of the experiment
is to integrate the e-mail and facsimile communities, providing a way for
e-mail users to send documents to the fax machines of people who do not

---

* The two people who started the remote printing experiment are Carl Malamud of
the Internet Multicasting Service, a nonprofit organization, and Marshall Rose of Dover
Beach Consulting.

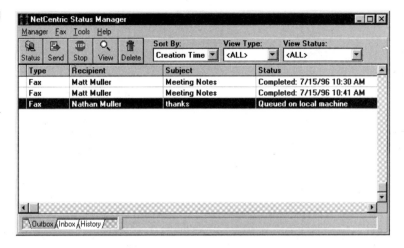

Figure 6.10
FaxStorm's Status
Manager window
provides a conve-
nient way to keep
track of all your faxes.

have e-mail. The arrangement is called *remote printing* because the remote fax machine prints out the document.

Working together, many servers cooperatively provide remote printing access to the international telephone network, allowing people to send faxes via e-mail to the calling areas of participating servers equipped with special fax spooling software. The general-purpose Internet e-mail infra-structure takes care of all the routing, delivering the document to the appropriate server (i.e., gateway) for distribution within a predefined call-ing area.

The remote printing facility is not available in as many locations as the commercial services and it is not intended for heavy usage. In fact, if a site administrator believes that a user is abusing the facility, further access can be denied. It is the responsibility of each Internet site running a remote printer server to define a local policy for denial of access. A remote

Figure 6.11
FaxStorm's Fax Detail
window shows all
information about
the fax transmission,
including modem
speed, job ID, and
transmit time.

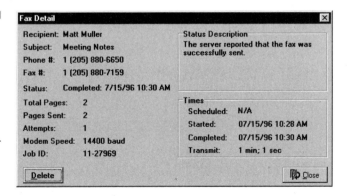

162

Figure 6.12
FaxStorm's Contact Manager lets you keep information organized by name, company, fax number, and phone number (the phone number column is not shown, but can be viewed on-screen by scrolling right).

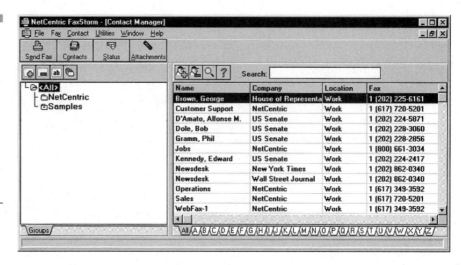

printer server might choose to impose a usage limit on a daily or monthly basis. Such limits are intended to balance the desire to encourage legitimate users with the need to prevent consistent abuse.

Commercial services have no usage restrictions; they simply bill subscribers accordingly. But the remote printing facility is free and requires no software other than your current e-mail program. Appendix A provides a list of locations, by country code and area code, served by the remote printing facility. You can request periodic updates by addressing e-mail to:

tpc-coverage@town.hall.org

The basic idea is that each participating site registers a *cell* indicating the portion of the international telephone number space to which it is willing to provide access. A cell can be a continent, a campus, a building, or a single phone number. There are four kinds of participating sites:

- Neighborhood sites
- Regional sites
- Enterprise sites
- Personal sites

A *neighborhood site* is a server that provides access to any facsimile machine in its local calling area. Since local calls add little or no extra charge to the site's phone bill, the site operator is willing to provide access out of a sense of community spirit. Access to Silicon Valley, for example, is provided by several neighborhood sites. Neighborhood sites can shrink or expand their cell, depending on such factors as demand and cost.

A *regional site* is just a large neighborhood site, usually providing access to an entire country or a large part of a country, such as an area code. The continent of Australia is an example of a regional site.

An *enterprise site* is run by a company that provides access solely to its own facsimile machines. It registers exactly those telephone prefixes which apply to its enterprise. The University of Michigan is an example of this. A geographically dispersed corporation or international association can also do this.

A *personal site* is run by someone who provides access to only his or her facsimile machine, usually one that resides on that person's own desktop. In this case, when the remote printer server gets the message, it will just deliver it to the owner of the desktop via e-mail.

To use the remote printing facility, you send electronic mail to an address which includes the phone number associated with the target facsimile device. Using the Domain Name System (DNS), the Internet message-handling infrastructure routes the message to a remote printer server, which provides access to facsimile devices within a specified range. The message is imaged on the target remote printer (i.e., fax machine) and an acknowledgment is sent back to the sender of the message via e-mail.

To send a document via the remote printer facility, you use the following format on the To: line of the message:

remote-printer.Joe_Smith@14157772525.iddd.tpc.int

You start the line by identifying the kind of access (remote-printer), along with the name of the recipient (Joe_Smith). Next comes the familiar @ sign, followed by the recipient's fax number. A dot separates the fax number from the acronym iddd, which stands for *International Direct Dialing Designator.* After the next dot comes tpc.int, which is the Internet subdomain.

With this information, the message will get routed to a remote printer server, which will transmit it as a fax to the recipient. When the transmission completes, an acknowledgment message will be sent back to you.

You can mix remote printing and e-mail recipients in the same message header, as in the following example:

To: remote-printer.Joe_Smith@14157772525.iddd.tpc.int

cc: Betty_Boop<bboop@ddx.com>

The Reply-to line of the message header can contain either a remote-printing recipient or an e-mail recipient.

Additional information can be included in the To: line of the message header, such as a room number that might expedite internal delivery at the remote location, as in the following example:

To: remote-printer.Joe_Smith/Room_206@14157772525.iddd.tpc.int

Notice that in all the examples, there are no spaces. This is because some mail programs have difficulty dealing with addresses that contain spaces. You must also be careful about what special characters you use to identify recipients. It is generally okay to use upper- and lowercase letters, numbers, and two special characters: underscore and slash (_ and /).

When a cover sheet is generated, the underscore will turn into a space and the slash will turn into an end-of-line sequence. So, given the address above, the cover sheet might start with

Please deliver this facsimile to:

Joe Smith

Room 206

When the fax is sent over the Internet to the fax machine at the other end, you will get back a delivery receipt that looks like this:

```
Date: Tue, 11 Jun 1996 22:13:42 -0400 (EDT)
From: PMDF-FAX transmission acknowledgment <postmaster@tink.com>
Subject: FAX successfully sent to Bukasa Tshilombo
To: nmuller@ddx.com

Transmission of your FAX message to Bukasa Tshilombo at 12125552859
has been completed.
Subject: Contract Negotiations
Date and time: Tue, 11 Jun 1996 22:13:42 -0400 (EDT)
Pages sent (including cover page): 2
Estimated duration of the phone call: 1 minute and 17 seconds

Your message was processed at a site serving the following United
States area codes: 212/718/800/917 and parts of 516. This site is
operated by New York Net. For questions specifically related to the
FAX service at this location, including problem investigations, you
can send mail to faxmaster@new-york.net.
```

If you try to send a document to a location not served by a remote printer server, you will get back an error notification via e-mail that looks like this:

```
To: nmuller@ddx.com
Subject: Your Fax to +1-2056663333 sent via TPC.INT
From: tpcadmin@info.tpc.int
Date: Tue, 11 Jun 1996 22:41:49 +0100
Sender: bin@info.tpc.int

In reference to your FAX sent on Tue, 11 Jun 1996 16:40:21 CDT,
regarding Contract Talks:
```

```
We regret to inform you that the phone number you attempted to
reach is not currently being served by a remote printer operator.
We hope to have coverage in this area—perhaps you know somebody who
could operate a remote printer server?

Regards,

Mr. Arlington Hewes    tpcadmin@info.tpc.int
The TPC.INT Subdomain  http://www.tpc.int/
```

If you would like to learn more about remote printing and stay informed of the latest developments, you can send a note to:

tpc-rp-request@aarnet.edu.au.

Corporate Fax Solutions

Although a complete discussion of corporate networks is beyond the scope of this book, you should be aware that there are enterprise-level e-mail/fax solutions available. Among these types of products is Connect²FAX from Infinite Technologies. Connect²FAX is a fax server that installs on a company's existing Connect²-based messaging engine. It supports all modems and it runs on NetWare and non-NetWare networks. Like other products mentioned throughout this chapter, the message text is scanned into an image format and becomes a fax cover sheet. Additional pages may be added by attaching text, PCX, or DCX files to the message. Connect²FAX takes care of the transmission and notifies you of successful and unsuccessful results. It can be installed on a company's messaging server and shares the modem with other applications.

Another fax server is available from NetManage. The Chameleon Fax Server provides both in- and outbound fax capabilities from the Chameleon Mail system. Since it is fully SMTP-based, the server can transmit faxes as well as mail over the Internet. And since it supports MAPI, you can fax documents right from within an application.

Menus and dialogs walk you through the setup, allowing you to create groups for group faxing, and configure the system for such things as number of retries, redial interval, and disk space threshold. Cover sheets can be created in Microsoft Word and saved at the server. The server also automatically converts file attachments to their native format.

The Chameleon Fax Server also comes with an application called the Coordinator, which permits you to specify the routing of incoming faxes if automatic routing—through direct inward dial (DID)—is not selected. The Coordinator presents you with a list of authorized users. After a new fax is received at the server, the cover page is then sent to the Coordinator. The Coordinator views the cover page and selects which recipients should receive the fax. This capability provides added security for confidential faxes, ensuring that only those authorized to receive faxes will get them. The server provides the sender with delivery notification.

The fax server operates over any Windows 3.1 or Windows 95 operating system. You need Chameleon Mail to send faxes from the server, but any mail client can receive faxes. Faxes can be opened with any viewer that can read PCX or TIFF files. The fax server is compatible with over 320 fax modem types and provides advanced configuration options for those who want more flexibility over the control of modems.

Windows Fax Software

If you're not interested in subscribing to a mail-to-fax gateway service or find the remote printer facility too limited, there are inexpensive stand-alone alternatives such as FAXfree from TAC Systems and JetMail Fax Server from NetManage.

TAC Systems offers a shareware and commercial version of FAXfree, which can be downloaded from its Web page at:

http://www.tacsys.com

The shareware copy of FAXfree allows you to send faxes containing graphics and text anywhere in the world over the Internet. All you need is the FAXfree software, Windows 95, and a connection to the Microsoft Network (MSN) or some other MAPI-compliant (Mail Application Programming Interface) service. At the destination end, the fax is received via e-mail as an attachment. The attachment is an image file that uses Microsoft's .AWD file extension. The attachment is opened and printed using Imaging for Windows 95.

Upon installing the FAXfree software, a FAXfree print driver will be put into your Printers folder in Windows 95 (Fig. 6.13). Although this print driver will allow you to fax documents from any application, Microsoft Word for Windows 7.0 is used here for discussion purposes. From within Word 7.0, you can select a fax cover sheet from the Template subdirectory. Upon selecting an appropriate cover sheet, save it from a .dot file to a .doc file. Now you can revise the .doc file as appropriate for your

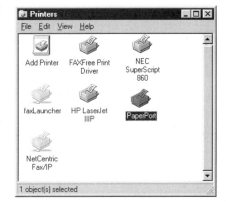

Figure 6.13

Upon installing the FAXfree software, the FAXfree print driver will be put into your Printers folder in Windows 95.

own needs and type a message in the space indicated (Fig. 6.14). The advantage of using the template is that the format is already laid out for you and it includes a clock function that displays the time.

After you have prepared the fax, select Print. From the list of printers, select FAXfree Print Driver. This will open up FAXfree's New Message window (Fig. 6.15). You just fill in the To:, Cc:, and Subject: fields as appropriate—being sure to enter an e-mail address in the To: field. In the large field in the middle of the window you can type in a brief message, which will

Figure 6.14

A fax cover sheet, selected from the Template subdirectory within Microsoft's Word for Windows 7.0, with appropriate header revisions and a message.

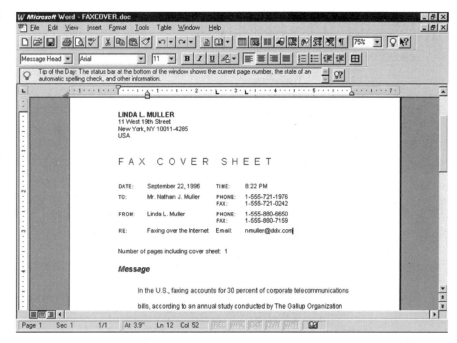

Figure 6.15
The New Message window in FAXfree. The object at the bottom is an image of the document prepared in Word for Windows 7.0, which will become the attachment to the e-mail message.

show up as ordinary text in the body of the e-mail message. In the last field at the bottom of the window, is an object. This object is an image of the fax (i.e., the Word 7.0 document) with a file name assigned by FAXfree and an .AWD extension. This will become the attachment to the e-mail message.

With the New Message window properly filled out, you are ready to send the fax by clicking the Send icon. This opens the TAC FAXfree window (Fig. 6.16) where a record of all your faxing activity can be found. At this point, the fax is sent automatically. However, if you are not connected to the Microsoft Network when you click on Send, FAXfree will invoke it automatically. After the connection is established the fax will go out over the Internet.

Figure 6.16
The TAC FAXfree window provides a record of all your fax activity.

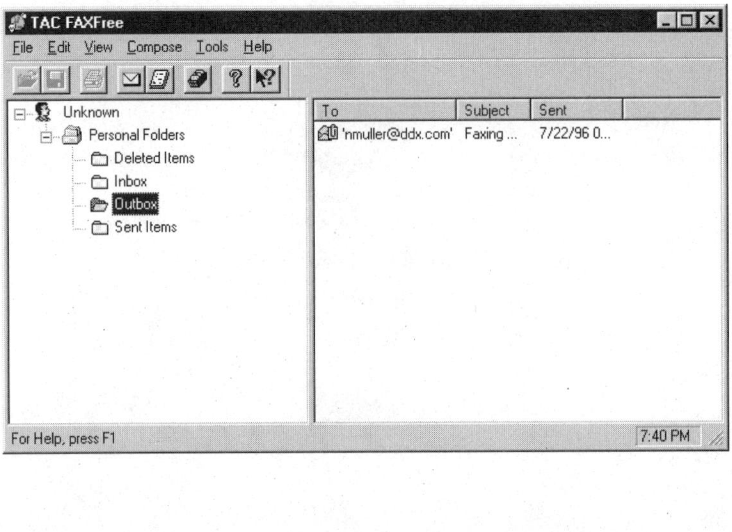

Figure 6.17

In the addressee's mail program—in this case, Eudora Pro— the last message in the list is tagged with a page icon, indicating the presence of an attachment.

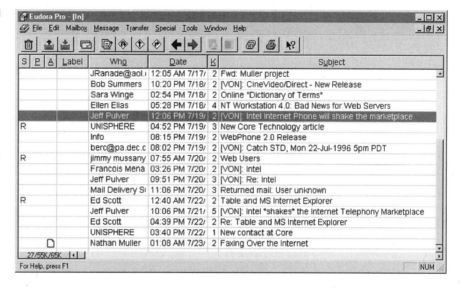

At the receiving end, the addressee opens his or her e-mail as usual. In the list of messages, the page icon next to the message at the bottom of the list indicates the presence of an attachment (Fig. 6.17).

When the addressee opens the e-mail, he or she will be able to read the brief message and see a notification at the bottom indicating where to find the attachment (Fig. 6.18).

The next step is to open the attachment, which is an image version of the original Word 7.0 document. Since the image has an .AWD extension,

Figure 6.18

The e-mail message, with an attachment notification at bottom, as shown in Eudora Pro.

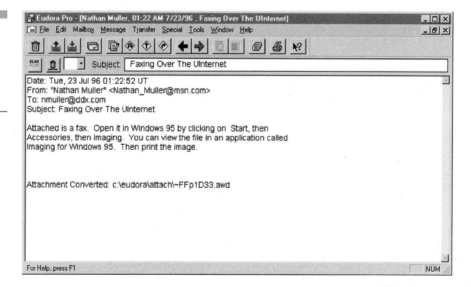

it must be opened with an image viewer that supports this type of file, such as Imaging for Windows 95. This is done by clicking Start, then Accessories, then Imaging. If your copy of Windows 95 did not come with this viewer, you can find it at Microsoft's Web site at:

http://www.microsoft.com

After opening the file with Imaging for Windows 95 (Fig. 6.19), you can save it, print it, and even annotate it before sending it to someone else. There are a few things to remember when using FAXfree:

- As the sender, you must use a MAPI-compliant service such as Microsoft Network; FAXfree will not work over such services as CompuServe, America Online, or other such third-party Internet gateways.

- The receiver is not restricted as to what service it uses to access the Internet. If you can send someone an e-mail message, you can send that party a fax via FAXfree regardless of the recipient's Internet connection.

- The recipient's mail system must be SMTP/POP3-compatible (i.e., Microsoft Exchange, PC Eudora, or similar mail program).

Figure 6.19
The image version of the original Word 7.0 document as viewed in Imaging for Windows 95.

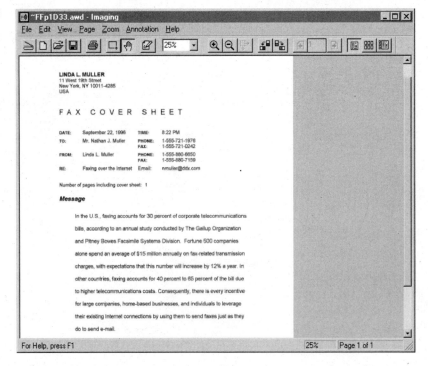

Although the evaluation version of FAXfree does not let you compose a fax cover page, you can include one as the first page in the attached document, as previously discussed.

The commercial version of FAXfree adds the following features:

- The ability to send a document to a conventional fax machine
- The ability to configure the program to redirect received faxes to another location
- Support for scanning, so you can print received faxes
- The ability to compose a fax cover page
- More extensive online help and product documentation

The use of FAXfree to send imaged documents to conventional fax machines and redirect faxes to other locations is intended for use over corporate intranets, which are private TCP/IP-based networks. However, FAXfree will not redirect fax documents if it results in a long-distance phone call. An e-mail message will notify you that the fax could not be sent for this reason.

Online Resources

The following table provides the Web links of the major Internet fax services and software providers from which you can download copies of their software and try out their services. The Web pages also contain such information as platforms supported, system requirements, product features, installation instructions, and troubleshooting advice. Since the technology is moving rapidly, and free offers surface periodically, you may want to access these Web pages for the latest developments.

Developer	Product	Web page or FTP site
DataSoft	DataSoft Message	http://www.data-soft.com/
FaxSav	FaxLauncher, FaxMailer	http://www.faxsav.com
Infinite Technologies	Connect²FAX	http://www.ihub.com/index.html
International Telecom	Faxaway	http://www.faxaway.com
IriSoft Works	MiEPO (My Electronic Post Office)	http://www.irisoft.be/vibuc/miepo.htm
JFAX Communications	JFAX Communicator	http://www.jfax.com

Developer	Product	Web page or FTP site
NetCentric Corp.	FaxStorm	http://www.netcentric.com
NetManage	Chameleon Fax Server	http://www@netmanage.com
Newsnet ITN	Faxscape	http://www.faxscape.com
TAC Systems	FAXfree	http://www.tacsys.com

NOTE: This information, as well as updates, can be found at http://www.ddx.com/mgh.shtml.

Conclusion

Cost savings is the principal reason for turning to fax solutions that work over the Internet. However, there could be reliability trade-offs to consider. Fax, like voice, has always been done in real time with a network connection that stays in place for the duration of the call. The Internet, on the other hand, is a connectionless packet-switching network that uses store-and-forward technology. This can result in significant delays in fax delivery, just as it sometimes does with e-mail. Transmissions can be delayed for a few hours, a few days, or even a week or more, depending on what kind of trouble the packets encounter along the way. Fortunately, the fax services described in this chapter will send you an e-mail message detailing any delivery problems so you can pursue other alternatives. Regular e-mail delivery subsystems on the Internet are not always this courteous. It may take a few hours or even the next day for a mailer daemon to inform you of nondelivery. Therefore, relying on the Internet may not be the best choice for sending high-priority business documents.

7

Paging Over the Internet

Introduction

The Internet continues to experience rapid growth. There are now about 40 million users on the Internet worldwide and that figure is growing by an estimated 10 percent per month. There are almost as many users of paging services—more than half of them in the United States—so it was almost preordained that there would someday be a marriage of the Internet with wireless paging networks, enabling Internet users to reach out and page someone who is not at his or her desktop computer to receive e-mail. Well, that day has come. If you have a Web page and subscribe to a wireless service that supports alphanumeric paging, you can escape from your corporate or home office and receive short messages to your pager wherever you happen to be—on vacation, enjoying a long weekend, riding in your car, or just putzing around the house. There are services that can handle this for you, or you can save on monthly add-on charges by linking to free mail-to-pager gateways on the Internet or by writing your own pager form that can be accessed from your Web site.

About Paging

A paging system provides one- or two-way wireless messaging to give mobile users continuous accessibility to family, friends, and business colleagues while they are away from telephones connected to the wired communications network. Typically, the mobile user carries a palm-sized device (the pager or some other portable device with a paging capability) which has a unique identification number. The calling party inputs this number, usually through the public telephone network, to the paging system which then signals the pager to alert the called party. The called party receives an audio or visual notification of the call, which includes a display of the phone number to call back. A mobile user can be alerted to important messages anytime, anyplace—even across national borders.

There are numerous practical benefits to subscribing to a paging service; among them are the following:

- *Improved productivity.*   A pager is one of the most cost-effective and reliable tools for improving productivity. It enables users to eliminate unnecessary, time-consuming travel as well as the costs associated with travel.

- *Competitive advantage.*   In today's business environment, using pagers can help keep a company in touch with customers, suppliers, strategic partners, and other important constituents, thereby conveying a responsive image. With employees carrying a pager, critical messages will be answered in a timely manner so that important business opportunities will not be missed.

- *Personal freedom.* Carrying a pager can give a user the freedom and flexibility to leave the office or home without giving anyone advanced notice or leaving an itinerary. Carrying a pager means that a person can always be reached, regardless of location or time of day.

- *Reduced stress.* A pager can prevent the irritation, frustration, and annoyance that comes from missing important calls, the inability to answer an urgent request, or not being notified of a changed appointment.

For these and other reasons, the pager has become very popular in recent years. It is also lightweight and small enough to fit into a shirt pocket, attach to a belt, or wear around your wrist like a watch. There are now a variety of alerting methods: vibration, flashing lights, and different beep tones. Advanced features include *group call,* which activates a number of pagers with a single call and is useful to emergency response teams. Another advanced feature is *message storage,* which permits review of saved phone numbers at an appropriate location or time. Many pagers now offer an alphanumeric capability that enables short messages to be received—usually no more than 240 characters—instead of just phone numbers.

The paging services themselves are becoming more advanced as well. Instead of being limited to one-way transmissions to the pager, there are now services that offer two-way transmission, allowing users to respond to pages with brief messages or stored responses. A service option that makes use of two-way transmission is called *display messaging.* This service allows callers to send three different types of messages to pagers and so-called *message-ready* cellular phones:

- A telephone number which acts like a digital page and signals a request for a callback

- A short alphanumeric message selected from a preprogrammed list of commonly used messages such as *call home* or *call office*

- A private message on voice mail

Users can scroll through their messages and, by pressing the appropriate button on a two-way pager or cellular phone, they can send a reply instantly without dialing the number. They can also choose to store messages and act on them in order of priority or read them at a more convenient time.

Because paging systems make efficient use of the radio spectrum—sending small amounts of information in short bursts—they can be used to deliver a variety of useful services at a relatively low cost. The introduction of complementary products—pocket-sized cellular phones, personal digital assistants, and notebook computers with wireless modem cards—has encouraged the paging industry to provide a range of value-added services. With alphanumeric capabilities has come a variety of new services that deliver financial news, stock quotes, sports scores, lottery results, and airline flight information in near real time. These devices can also receive short-text messages via e-mail originating from the Internet.

Paging Networks

All paging networks—whether on-site, citywide, regional, or nationwide—are based on the same technology. A radio transmitter, installed on a ground-based tower, broadcasts a constant stream of messages on a specific radio frequency (or channel). The receivers (i.e., pagers) are tuned to that channel and listen to the message stream, waiting for messages intended for their unique address. When the receiver detects a message tagged with its address, it notifies the subscriber, who can view the phone number or message on the display.

The original pagers, or beepers, received only the address itself from the transmitter and emitted a beep, alerting the subscriber to call his or her office to pick up a message. In the 1970s, voice circuitry was added to pagers, resulting in a hybrid unit known as a *tone-voice pager.* This type of pager signals an incoming message with a tone and then replays a short message dictated by the caller. While these units are popular in select commercial markets, the amount of valuable air-broadcast time required for a typical voice message makes their cost prohibitive for personal use. This will change in the near future as *Personal Communications Services* (PCS) are introduced. This category of services will have enough bandwidth to deliver voice as well as data to handheld devices such as Personal Digital Assistants (PDAs) and hybrid cellular phones.

The next innovation—*digital display paging*—was introduced in the early 1980s. With this type of pager a caller enters a callback number through a standard telephone keypad and that number is displayed on the subscriber's pager.

In the late 1980s, *alphanumeric pagers* were introduced. With these units, actual short-text messages can be entered from a computer and displayed on a pager's screen. Because these pagers often eliminate the need to call the sender back to obtain more information, they have become very popular.

E-mail versus Paging

Many organizations already use the Internet to connect e-mail systems with gateways based on the Simple Mail Transport Protocol (SMTP). Through these gateways, short-text messages can even be sent to pagers via any kind of e-mail software. The drawback of SMTP is that it doesn't provide immediate notification or message acknowledgment, which is the essence of paging.

With a separate gateway using a protocol known as the *Simple Network Paging Protocol* (SNPP), which has been around since 1994, a user connected to the Internet can page a mobile recipient with an urgent message and receive an acknowledgment from the paging network. The message is transmitted through the Internet to a paging network via the SNPP-equipped gateway, and then to any alphanumeric paging device, including a laptop computer equipped with the appropriate PCMCIA card.

SMTP works well for traditional messages, but by tying the SNPP gateway to the Internet, users can be assured that messages will be sent, received, and confirmed immediately—usually within one minute. If a message does not get delivered, users would not receive an acknowledgment and would be able to use alternative communication methods.

E-mail doesn't work that way. A good example of this is deferred messaging, which happens when a gateway is down. Suppose Linda Tyke (*ltyke@newventure.com*) sends a message to Robin Muller's pager (*5551212 @pager.pagingcompany.net*). However, Linda's gateway to the Internet is down, causing her message to be deferred. Linda is not notified of this delay because her message has not actually failed to reach its destination. Three hours later, the link is restored, and the message is sent. If Linda's page concerned an important meeting that was supposed to happen two hours ago, the information obviously did not arrive in time to be useful. On the other hand, if Linda had used her SNPP client software, she would have immediately discovered the network problem. She could then have used the telephone to call Robin's pager.

The issue is not page delivery, but the immediate notification of a problem that affects your message. Standard e-mail and SMTP, while quite reliable in most cases, cannot guarantee timely delivery between all nodes at all times, making it less desirable for emergency or urgent messages.

Paging Services

Among the many paging services that allow you to send messages via the Internet is PageMart. PageMart is a nationwide paging system, covering 1100 major U.S. communities in all 50 states. PageMart supports text paging over the Internet using SMTP or SNPP.

Senders use an address consisting of the Pager Identification Number (PIN) assigned to the subscriber's text pager, followed by the domain name *pagemart.net*. For the PIN 1234567, for example, the e-mail address would be *1234567@pagemart.net*. PageMart subscribers can even set up their own forwarding rules so urgent e-mail messages received at their normal Internet, corporate, or e-mail service mailbox will reach them as soon as possible via pager.

More information about PageMart and its services is available from the company's Web site at:

http://www.pic.net/business/pagemart/pagemart.html

Another company, Mtel Corporation, provides paging to more than a million customers worldwide under its SkyTel brand. From its Web page at:

http://www.skytel.com/paging

you can send messages to subscribers of the company's SkyWord and SkyTel 2-Way services (Fig. 7.1). Messages can be 240 characters for SkyWord subscribers and 500 characters for SkyTel 2-Way subscribers. You can even create custom replies (up to nine) that the recipient can respond with. Messages from Internet users reach SkyTel pagers via the SkyTel Internet gateway.

Alternatively, SkyTel 2-Way messages can be created and sent through any Internet-based e-mail system. Mtel also offers SkyTel QuickAccess, which is paging software anyone in the United States can use on a modem-equipped PC or Macintosh to send messages to SkyWord and SkyTel 2-Way subscribers. The software can be downloaded from the SkyTel Web site at:

http://www.skytel.com/products/skyword.html#quick

Figure 7.1

Callers can send messages to SkyWord and SkyTel 2-Way subscribers using this Web form. All you need to send a message is the person's SkyWord PIN or SkyTel 2-way mailbox ID. Replies will be directed back to the e-mail address. (Some SkyWord customers may have to add e-mail paging to their account for this service to work.)

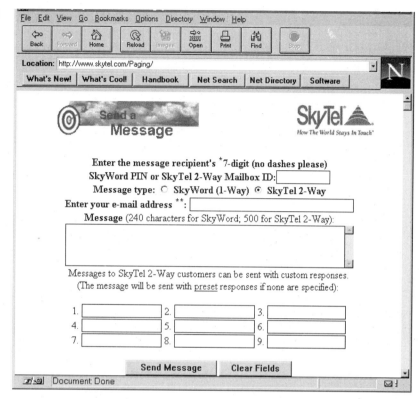

SkyTel Access is another software package that lets you compose and send messages directly from your PC or Macintosh computer to anyone on the SkyTel network without going through the SkyTel operator. An address book provides easy access to the PINs or Mailbox IDs of the people you need to contact. The software offers more functionality to SkyTel 2-Way users in that they can receive a message with multiple reply choices and respond appropriately by pressing a button. When composing a message, you can create your own custom replies to fit your message, or you can allow the recipient to respond with any of 16 preset replies:

- YES/OK
- NO
- WILL CALL LATER
- CALL ME
- RUNNING LATE
- NEED MORE INFO

- SEND # TO CALL
- WHERE ARE YOU?
- WILL ARRIVE 15M
- WILL ARRIVE 30M
- TRAFFIC DELAY
- PICK ME UP
- BUSY
- FINISHED
- CALL HOME

SkyTel Access provides an outbox so you can track the delivery of Sky-Tel 2-Way messages. You can retrieve SkyTel 2-Way replies e-mailed to you directly from your computer. An inbox lets you retrieve any messages you've received over a specified period of time. From here, you can send a reply, forward the message to another user, or save it to a text file on your PC. The SkyTel Access software is free to anyone in the United States, but at this writing was not available for download from the SkyTel Web site.

PageNet, one of the largest wireless messaging companies in the world, has more than 7 million subscribers in the United States alone. PageNet's two-way alphanumeric paging service allows you to receive messages of up to 250 characters per message. PageNet allows you to receive e-mail, fax, and data messages on your pager, each time- and date-stamped to let you know when the message was received. With the PageMail feature, your pager alerts you when the caller leaves a recorded phone message, so you can retrieve it by Touch-Tone phone.

You can send a page to any PageNet subscriber by accessing the company's Web pager (Fig. 7.2) at:

http://www.netpage.com

MCI offers a Web pager (Fig. 7.3) that lets you send a page to anyone who has a networkMCI or SkyTel one-way pager (networkMCI Interactive and SkyTel 2-Way paging are not yet supported). All you need is the pager's PIN and you can send a 240-character message to alphanumeric pagers or 10 characters to a numeric-only pager. MCI's Web pager is located at:

https://cons.mci.com/paging/textsendpage.shtml

Figure 7.2

At this writing, PageNet's Send A Page capability is in the beta test stage. The company is in the process of building a comprehensive Internet paging service. To send a page, the sender only needs to know the paging subscriber's nationwide Pager Identification Number (PIN).

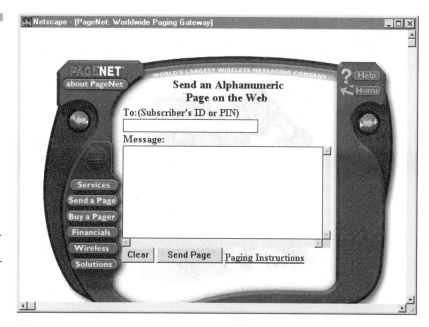

Figure 7.3

MCI's Send a Page. In addition to sending a page, you can also send your message to an e-mail address.

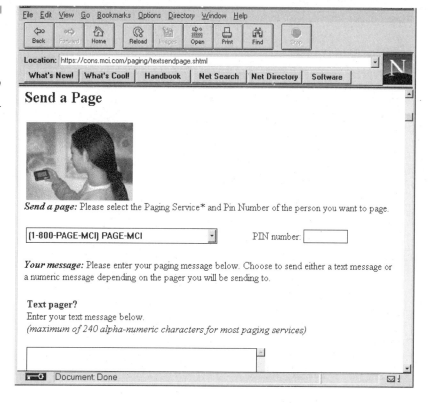

Web Pagers

There are several places on the Web where you can send messages to pagers. At Information Architecture's Web-Page, for example, you can send pages directly from a Web-based form to anyone who has a networkMCI or SkyTel pager. The form (Fig. 7.4) is located at the following Web site:

http://www.iarc.com/ia/code/pager2.html

You just select the type of pager service from the pull-down list of eight services and enter the PIN of the person you want to page. The message length is limited according to the pager you are sending the message to, which in most cases is no more than 240 characters. Anything longer is cut off when the message is received or rejected entirely. You also can send the same pager message to the recipient's e-mail address.

Figure 7.4

From Information Architecture's Web-Page, you can send messages to anyone with a networkMCI or SkyTel pager.

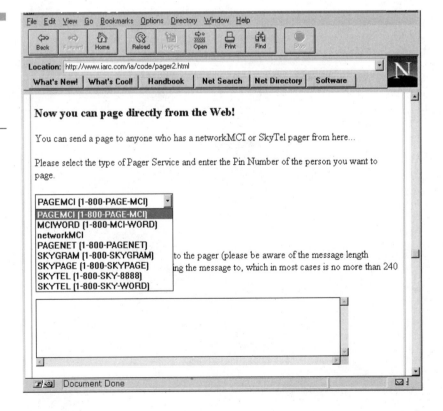

Figure 7.5

Keith Gabryelski's Web Pager, hosted by Wildfire Communications, allows you to send messages to subscribers of about 40 different paging services.

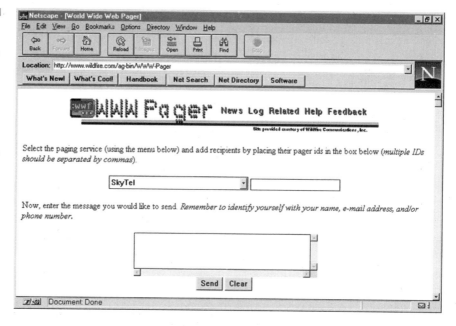

Wildfire Communications hosts a Web form written by Keith Gabryelski that can be used to send messages to subscribers of about 40 different paging services (Fig. 7.5). This form can be accessed at:

http://www.wildfire.com/ag-bin/WWW-Pager

Also included at this site are directions for setting up a link from your home page to this service and automatically specifying you as the recipient of the messages.

Other Internet Pagers

Pagers can be used for a variety of other Internet-related applications. For example, TVision offers a software product called WebPager, which is used to monitor the performance of your Web site and notify you via pager when your site fails to respond to visitor accesses. It supports single or multiple Web sites.

WebPager provides functionality to monitor the status of a collection of URLs (Web documents or document components). The program will visit each URL on a user-selected schedule. If a URL does not respond to a request, WebPager will send a message to a pager specified by the user. One does not need root privileges to run WebPager.

Pages are sent over the Internet using the Simple Network Paging Protocol (SNPP). As previously discussed, the advantage of using SNPP to send a page, versus the more commonly used practice of sending an e-mail message to a pager site, is that the speed of transmission is on the order of minutes. An e-mail message, on the other hand, can take up to several hours to travel through a corporate gateway or over a series of mail hosts. Furthermore, when sending an e-mail message, there is no confirmation by the pager site that the message made it through at all. With SNPP, the pager server sends an immediate confirmation that the paging data has been received.

By combining the SNPP protocol with HTTP client capabilities, WebPager provides a more reliable way of ensuring that a Webmaster or technician is quickly notified of any site failure.

WebPager can be configured to invoke specially constructed CGI programs at your site. This capability can be used to perform

- Disk space monitoring
- Processor performance monitoring
- Network performance monitoring

Additional information about the WebPager can be found at the company's Web site at:

http://www.tvisions.com/index.html

The Internet Appliance

A relatively new type of device is the AT&T PocketNet Phone from AT&T Wireless Services. The PocketNet Phone is an integrated cellular phone and wireless Internet appliance that provides people on the go with fast and convenient access to Internet information and two-way messaging services, including e-mail, airline flight information, and financial information. An array of personal services will become available, including sports scores, local movie listings, and lottery results.

At the heart of the AT&T PocketNet Phone is a specialized browser that is specifically tuned to send and retrieve only text-based information that resides on the Internet, not burdensome multimedia and graphics. Browser and server applications are written in the Handheld

Device Markup Language (HDML), an open programming specification derived from the Internet standard HyperText Markup Language (HTML). HDML allows applications to run on the existing Internet infrastructure without modification. With this approach, the browser optimizes the cellular phone's compact display size and limited memory for information services, while being fully compatible with existing Web applications.

The PocketNet Phone sends and receives information via AT&T's wireless IP network, also known as Cellular Digital Packet Data (CDPD). Since CDPD is TCP/IP implemented over the cellular network, all Internet-based applications work over CDPD without modification. This allows Web developers to use existing network applications and tools when designing for the wireless environment.

To access information or receive e-mail messages from the Internet, the user manipulates the browser's menu-based user interface with the cellular phone keypad. Requests for information are routed through the wireless IP network and the wireline Internet (or a proprietary wide area network) and processed at the Web server on which the application resides. The browser displays only the results of the query or the e-mail message, four lines at a time, leaving out extraneous information.

Write Your Own Web Pager

You can subscribe to a commercial online service such as NetPage which allows you to receive pages from the Internet in one of two ways: from a simple form on your Web page or by e-mail sent to a special mail-to-pager account. More information about the NetPage service is available at its Web site at:

http://www.net-pager.com/

If you have your own Web page, you can create hypertext links to the various no-fee Web pagers, allowing visitors to reach you when you are not sitting in front of your computer. However, you can create your own Web pager with the advantage of maintaining complete control. You can write your own Web pager form in HTML and process it with a Common Gateway Interface (CGI) script written in Perl to send it to your pager. Of course, for this to work, you must subscribe to a paging service and have a pager capable of receiving text so that others can use this form to reach you.

For this demonstration, we will write an appropriate HTML form and a Perl script to process and deliver messages to a PageMart pager. When the user hits the Send button on the form, an appropriate acknowledgment will be displayed, repeating back his or her name and phone number. This lets the sender know that the page is being handled, rather than leaving him or her guessing about whether the form worked. Repeating back the sender's name and phone number adds a nice personal touch. We will also have the Perl script send a copy of the message back to the sender's e-mail system so he or she can have a permanent copy of the page for future reference.

The HTML Form

The first item you will need is an appropriate HTML form. This form uses the same tags used to build the Email Notification Service form described in Chap. 4, including the POST method of delivery, as specified in the first FORM tag. The FORM tag also references the location and name of the Perl file as "/nmullerbin/pager.pl". As explained in Chap. 4, nmullerbin is a bin alias that is set up at the Web server, while pager.pl is the name of the Perl script that supports this form (discussed later). The bin alias is merely shorthand for a much longer directory/subdirectory name which identifies the exact location of the Perl script on the server. The TEXTAREA tags define the field into which the user will enter the message that will be delivered to the pager upon hitting the Send button.

The following HTML code will get you started, but you can customize it any way you like:

```
<HTML>
<HEAD>
<TITLE>My Web Pager</TITLE>
</HEAD>
<BODY>
<P>
<CENTER><H2>My Web Pager</H2></CENTER>
<P>
<HR=5>
<P>
<CENTER>You can reach me anytime via pager. Do not use more than
240 characters in the message field.</CENTER>
<P>
<form method="POST" action="/nmullerbin/pager.pl">
<B>Name:</B><BR>
<input TYPE="text" NAME="name" size=40>
<P>
<B>Organization:</B><BR>
```

```
<input type="text" NAME="organization" size=40>
<P>
<B>Email Address:</B><BR>
<input type="text" NAME="e-mail" size=40>
<P>
<B>Phone Number</B><BR>
<input type="text" NAME="phone" size=40>
<P>
<B>Message:</B><BR>
<textarea name="message" ROWS=8 COLS=40></textarea>
<P>
<CENTER>
<input type=submit value="Send">
<input type=Reset value="Clear">
</CENTER>
</FORM>
<P>
<HR>
<P>
<A HREF="index. html"><IMG SRC="al03.gif" ALIGN=MIDDLE></A><EM> Go
back to Main Menu</EM></P>
</BODY>
</HTML>
```

The results of this HTML coding are shown in Fig. 7.6.

The Perl Script

The following Perl script is used to process the inputs from the HTML form. The script is divided into sections that describe the functions of the next lines of code. These comments are preceded by the pound sign (#), which is ignored by the Perl interpreter. You can use the script as is or customize it any way you like. The script is as follows:

```
#!/usr/local/bin/perl

$sendmail = "/usr/lib/sendmail";
$to = '1234567@pagemart.net, nmuller@ddx.com';

#1 ---- Sets up the date-time stamp for the On-screen Acknowledgment
message ---

($sec, $min, $hour, $mday, $mon, $year, $wday, $yday, $isdst) =
localtime(time); %weekday = ("0", "Sunday","1", "Monday","2",
"Tuesday","3", "Wednesday","4", "Thursday","5", "Friday","6",
"Saturday",);

%month = ("0", "January","1", "February","2", "March","3",
"April","4", "May","5", "June","6", "July","7", "August","8",
"September","9", "October","10", "November","11", "December",);
```

Figure 7.6
The HTML tags are
used to create My
Web Pager form, as
rendered in Netscape
Navigator.

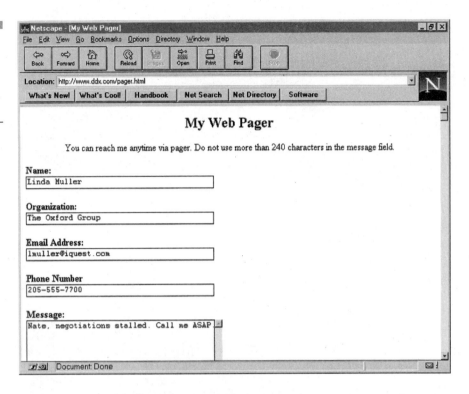

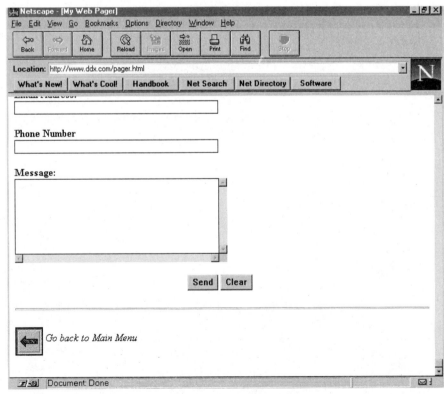

```
#2 ----- Sets up relationships of the variables -----
print "Content-type: text/html\n\n";
read(STDIN, $buffer, $ENV{'CONTENT_LENGTH'});
@pairs = split(/&/, $buffer);
foreach $pair (@pairs)
{
    ($name, $value) = split(/=/, $pair);
    $value =~tr/+//;
    $value =~s/%([a-fA-F0-9][a-fA-F0-9])/pack("C", hex($1))/eg;
    $value =~s/~!/~!/g;
    $FORM{$name} = $value;
}


#3 ---- Checks that the e-mail field on the HTML form is structured
properly ----
if ($FORM{'e-mail'} !~/.+\@[\w.]+/) {
print "<H2>This form is not complete.</H2><P>\n";
print "Please go back and provide an e-mail address.<P><HR><P>\n";
print "<A HREF=\"http://www.ddx.com/pager.html\"><IMG
SRC=\"http://www.ddx.com/al03.gif\" ALIGN=MIDDLE></A><EM> Return to
My Web Pager\n";
exit;
}


#4 ----- Checks that the phone number field in the HTML form is
filled in -----
if (length($FORM{'phone'}) <10) {
print "<H2>This form is not complete.</H2><P>\n";
print "Please go back and provide a phone number, including area
code.<P><HR><P>\n";
print "<A HREF=\"http://www.ddx.com/pager.html\"> <IMG
SRC=\"http://www.ddx.com/al03.gif/\" ALIGN=MIDDLE></A><EM> Return
to My Web Pager\n";
exit;
}


#5 ----- Checks that the message field on the HTML form is filled
in -----
if ($FORM{'message'} eq "") {
print "<H2>This form is not complete.</H2><P>\n";
print "Please go back and enter a message.<P><HR><P>\n";
print "<A HREF=\"http://www.ddx.com/pager.html\"> <IMG
SRC=\"http://www.ddx.com/al03.gif\" ALIGN=MIDDLE></A><EM> Return to
My Web Pager\n";
exit;
}


#6 --- Checks that the message field does not contain more than 240
characters ----
if (length($FORM{'message'} )>240 ) {
print "<H2>Sorry, there are too many characters in the message
field.</H2><P>\n";
print "Please go back and shorten your message.<P><HR><P>\n";
print "<A HREF=\"http://www.ddx.com/pager.html\"> <IMG
SRC=\"http://www.ddx.com/al03.gif\" ALIGN=MIDDLE></A><EM> Return to
My Web Pager\n";
exit;
}
```

```
#7 ----- Formats the Page ------
open (SMAIL, "|$sendmail $to") || die "Can't open $sendmail!\n";
print SMAIL "To: Nathan Muller\n";
print SMAIL "From: $FORM{'e-mail'} ($FORM{'name'})\n";
print SMAIL "Subject: My Web Pager\n";
print SMAIL "-----------------------------------------------\n";
print SMAIL "Organization: $FORM{'organization'}\n";
print SMAIL "Phone number: $FORM{'phone'}\n";
print SMAIL "-----------------------------------------------\n";
print SMAIL "Message: $FORM{'message'}\n";
close (SMAIL);


#8 ---- Formats a date-time stamped On-screen Acknowledgment to the
sender ----
print "<Head><Title>My Web Pager</Title></Head>";
print "<H3>Your page was sent:</H3>";
print "$weekday{$wday}<BR>";
print "$month{$mon} $mday, 19$year<BR>";
print "$hour:$min:$sec CST";
print "<HR>";
print "<P>\n";
print "<BODY><H2>Thank you $FORM{'name'}.</H2><P>";
print "<H3>I will call you at $FORM{'phone'} as soon as
possible.</H3></BODY><P>\n";
print "<HR><P>\n";
print "<A HREF=\"http://www.ddx.com/index.shtml\"> <IMG
SRC=\"http://www.ddx.com/al03.gif\" ALIGN=MIDDLE></A><EM> Return to
Main Menu</EM></FONT>\n";


#9 ----- Sends a Pager Receipt to the sender via e-mail (SMTP) -----
$sendmail = "/usr/lib/sendmail";
$to = "$FORM{'e-mail'}";


open (SMAIL, "|$sendmail $to") || die "Can't open $sendmail!\n";
print SMAIL "From: Nathan Muller\n";
print SMAIL "Subject: Pager Receipt\n\n";
print SMAIL "You sent the following page to Nathan Muller: \n";
print SMAIL "\n";
print SMAIL "$FORM{'message'}\n";
close (SMAIL);
```

The first line of the script tells the server to run the script through the Perl interpreter. The program will not work without this line. In addition, this line must be followed by a blank line.

The next line,

```
$to = '1234567@pagemart.net, nmuller@ddx.com';
```

contains your PIN number (if you are a PageMart subscriber) and your e-mail address, indicating where the form will be delivered. In this case, the message will be delivered to both your pager and e-mail system. The two addresses are separated by a comma.

Section 1 of the script sets up a date-time stamp that will be used in the on-screen acknowledgment. It will tell the sender the exact date and time the message was sent—not the time the message was delivered. The time of delivery will vary, depending on network conditions.

Section 2 of the script specifies the content type; in this case, text and/or HTML-coded text. After that, the next lines in the program establish the relationships of name and value pairs.

Section 3 of the script does pattern matching on the e-mail address supplied by the sender to ensure it is present and, if so, properly structured. It must have an @ sign, followed by a character and a period. If not, an error message will be displayed when the user tries to send the form (Fig. 7.7).

Section 4 of the script checks that the sender's phone number is filled in. If so, it must contain at least 10 characters, including area code. If not, an error message will be displayed when the user tries to send the form.

Section 5 of the script checks that the message field in the HTML form is filled in. If it is empty, an error message will be displayed when the user tries to send the form.

Section 6 of the script checks that the message field does not contain more than 240 characters. You can change this number to suit your par-

Figure 7.7
If the e-mail address is missing from the HTML form or improperly structured, the Perl script sends back an error message telling the user to provide an e-mail address.

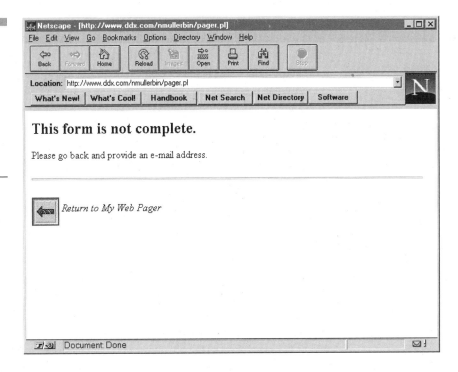

ticular paging service. Some will handle messages of 500 characters or more.

Section 7 of the script sets up the various parts of the message for display on the pager's screen and e-mail client.

Section 8 of the script formats the date-time stamped acknowledgment message that will be displayed on the user's screen after sending the page. The date-time stamp shows when the message was sent, not when it was delivered.

Section 9 of the script sends a pager receipt back to the sender's e-mail address. This is intended to provide the sender with a permanent record of the page. It is not a confirmation of delivery. Confirmation of delivery is provided separately, if the service provider's mail-to-page gateway on the Internet supports the Simple Network Paging Protocol (SNPP).

Initializing the Script

As explained in Chap. 4 in the discussion of how to create your own Email Notification Service form, before you can use this Perl script, you must put it into a cgi-bin subdirectory on the server. If you do not already have a cgi-bin directory, you must either establish a Telnet session with the server to create one or request that the system administrator create it for you. The HTML-coded forms can be located in any directory, preferably with your other HTML documents. HTML documents and Perl scripts are loaded into their respective directories via FTP.

Once the Perl scripts are placed in the cgi-bin subdirectory in ASCII form (never save them in binary form), they must be initialized so they will be executable. You can have the system administrator do this for you, or you can do it yourself via a Telnet session. After loading the Perl script to the cgi-bin subdirectory via FTP, just access the directory via Telnet and perform the following function (assuming a UNIX server) the first time the Perl script is loaded:

```
chmod a+rx pager.pl
```

This determines who gets to use the file for processing forms. Essentially, *chmod* means change mode, which makes the script executable and sets the permissions; specifically, this allows *pager.pl* to read (r) and execute (x) for everyone (a+). You only have to do this procedure the first time you load a Perl script. You can modify and reload a script any number of times via FTP and the permissions will hold unless specifically changed. Any time you modify and reload the Perl script via FTP, it must be as an ASCII file and not binary. Uploading the file as binary will introduce hard returns into the Perl script which will render the script unusable.

Pager Output

The Perl script will send you a short-text message on your pager that looks something like this:

```
Date: Fri, 2 Aug 1996 23:48:15 -0500
To: Nathan Muller
From: lmuller@iquest.com (Linda Muller)
Subject: My Web Pager
--------------------------------------------
Organization: The Oxford Group
Phone number: 205-555-7700
--------------------------------------------
Message: Nate, negotiations stalled. Call me ASAP.
```

On-Screen Acknowledgment

The last lines in the Perl script, which begin with the print command, will cause an appropriate acknowledgment message (Fig. 7.8) to be dis-

Figure 7.8
The acknowledgment message as rendered in Netscape Navigator.

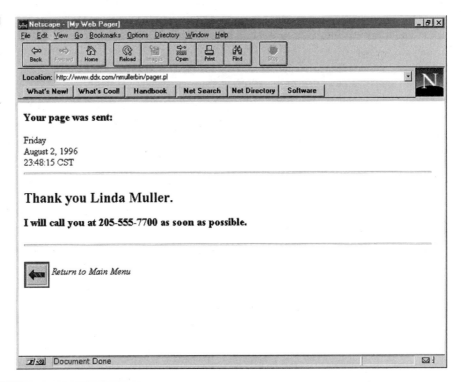

played to the user after the Send button is clicked. As noted, this lets the user know that the page is being handled, rather than leaving him or her guessing about whether the form worked.

Pager Receipt

The Perl script will send back a short-text message to the sender's e-mail address. The pager receipt will look something like this:

```
Date: Fri, 2 Aug 1996 23:48:15 -0500
From: Nathan.Muller
Subject: Pager Receipt
Apparently-To: lmuller@iquest.com


You sent the following page to Nathan Muller:
Nate, negotiations stalled. Call me ASAP.
```

Online Resources

The following table provides links to the Web pages or FTP sites of the various Web pager services. Like other Internet-related technologies, Web

Developer	Product or Service	Web page or FTP site
DataSoft	DataSoft Message	http://www.data-soft.com/
Ex Machina	Web-Based Paging	http://exmachina.com/paging.shtml
Keith Gabryelski	WWW Pager	http://www.wildfire.com/ag-bin/ WWW-Pager
Galaxy Star Systems	Online Pager/Fax Center	http://www.galstar.com/newtech/page.html
Information Architecture	Web-Page	http://www.iarc.com/ia/code/pager2.html
MCI Communications Corp.	Reach Out from the Web With MCI	https://cons.mci.com/paging/textsendpage. shtml
Mtel Corp.	SkyNet	http://www.skytel.com/paging
NetPAGE	NetPAGE	http://www.net-pager.com/
PageNet	PageNet	http://www.pagenet.com
Sprint	Send A Page Online	http://www.sprint.com

NOTE: This information, as well as updates, can be found at http://www.ddx.com/mgh.shtml.

paging is moving rapidly, so you may want to consult these locations periodically for the latest developments.

Conclusion

As advances continue in mobile computing technology and interconnection, an increasing percentage of the information transmitted through paging networks is coming from computerized sources, rather than through telephone numeric keypad entry. This kind of transmission is accomplished through a mobile data transceiver which connects directly to a desktop computer's serial port or into a notebook computer's PCMCIA expansion slot, allowing any type of data to be sent directly to roaming devices. This opens up paging networks to such applications as spreadsheet and calendar updates, faxes, word processing documents, and any other type of information that a user might want to transmit. Other value-added services will be added, such as the ability to send an e-mail attachment consisting of a voice message or video file that can be played back by the handheld device. The next generation of networks that will handle these types of applications are already being deployed. These digital networks will support a new category of services known as Personal Communications Services (PCS).

In 1993, the federal government authorized the Federal Communications Commission (FCC) to auction off radio spectrum for two types of PCS networks: narrowband and broadband. Narrowband PCS will be used to provide such new services as advanced voice paging, two-way acknowledgment paging, and data services, while broadband PCS will be used to provide a variety of new mobile services to an emerging category of communications devices that will include small, lightweight multifunction portable phones, portable facsimile and other imaging devices, multichannel cordless phones, and advanced paging devices with two-way data capabilities. From 1994 to 1996, the FCC raised over $9 billion from these auctions. Since service providers had to pay for the rights to use the radio spectrum, they will have every incentive to build PCS networks and offer services to consumers quickly in order to recover their investments.

Broadcasting Over the Internet

Introduction

With an Internet connection and the right software, you can produce audio or video programs that can run live over the Internet on a scheduled basis or embed these programs into your Web pages using HTML tags for on-demand access. You can produce all kinds of programs: news, entertainment, education, and information. Heck, you can combine all of these and put out your own infomercial. Much of the hardware and software you need for broadcasting audio and video content over the Internet has already been discussed in previous chapters. The hardware—modem, sound board, speakers or headset, microphone, video camera—is inexpensive and the software is available from a variety of vendor Web sites as freeware or trialware.

Live performances are usually made available on a scheduled basis, whereas recorded programs offer the convenience of on-demand access. In both cases, communication is one-way. Although the technology exists for two-way communication, as in audio- or videoconferencing, the term *broadcast* implies one-way communication only. Programs can be virtually any length, from a few minutes to an hour or more. A recorded announcement of a new product or a political opinion, for example, might take only a few minutes and be embedded in your Web page as a hypertext link for on-demand access to the audio or video file, whereas a live performance of a rock group or a radio show might be an hour long and require that viewers or listeners access a specially equipped server from which the Internet broadcast emanates.

If your content is good enough, you can potentially reach a vast global audience of online listeners or viewers. Keep in mind, however, that you are competing not only with traditional radio and television—plus all the other forms of entertainment that tend to occupy people on a daily basis—but with some very professional people and organizations that are dedicated to the production of programs specifically created for the Internet. Some organizations, such as AudioNet, offer extensive broadcast offerings that include selections from new CDs, radio programs from a variety of domestic and international stations, concerts, news, special events, sports, and even audio books. In only one year of operation from 1995 to 1996, AudioNet had accumulated over 6000 hours of archived programming, which Web users can access and listen to on demand. You can get a sense of what professional Internet broadcasting is all about by visiting the company's Web site at:

http://www.audionet.com

There are a number of examples of how you can use Internet-based audio or video broadcasting. For example, if you are a comedian, singer, musician, or in a band (or any other type of entertainer), you can broadcast live or recorded performances and reach an audience you would never have had the chance of reaching through any other means. It doesn't require a booking agent, public relations firm, auditions, travel, or a contract with a big-name recording studio or record label. If you have your own Web page, you can even provide an online order form for the purchase of tapes and compact discs (CDs) of your collected works.* If your performance is good enough, your work may even get into the hands of key players in the entertainment business and open up all kinds of new opportunities to advance your career.

If you are a newsletter publisher, you can broadcast samples of your content over the Internet with the idea of attracting new subscribers. Broadcasting is especially well suited for this purpose because you don't have to give away samples of your newsletter in the hope of increasing subscribership. This does away with associated printing and mailing costs for extra promotional copies, plus you do not have to rent mailing lists which are usually limited to one-time use. Alternatively, you can offer subscribers free access to your Web page to get up-to-the-minute audio or video reports that supplement newsletter content. This can be promoted as a value-added service to encourage new subscriptions.

If you are a radio broadcaster, you can reach pockets of loyal listeners who may have moved to other parts of the country and miss hearing about their favorite sports teams or radio personalities. Alternatively, you can get more mileage out of your production dollars by making taped broadcasts available over the Internet for on-demand listening. You can even offer uncensored versions of the material that may have been cut before broadcast. Archiving shows so listeners can selectively listen to them at a more convenient time can also present new opportunities for revenue generation from advertisers. The radio station can have a marketing edge in being able to show potential advertisers how their message lives on past the original 15- or 30-second spot.

If you are active in politics, social causes, charities, or community events, you can offer audio or video content over the Internet for such purposes as soliciting donations, recruiting volunteers, keeping members

---

* The technology for mastering and replicating your own CDs, including professional-quality labels, is now fairly inexpensive, costing well below $2000. However, further discussion of this topic is beyond the scope of this book.

informed of issues, and mobilizing them for action. While text-based Web pages are often used for these purposes, audio or video content adds more impact and urgency to the message. In turn, these messages tend to be retained longer by listeners or viewers and are more likely to be acted upon when received.* You can even produce audio or video documentaries about what your organization is all about, what it seeks to accomplish, and how the involvement of "people like you" can make a difference.

If you are an educator, you can tape your lessons and make them available over the Internet. This gives students a valuable reference while doing homework, reviewing assignments, making up lost ground due to absence, or preparing for important tests. Via live broadcast, you can also extend the reach of certain courses to students at other schools (or corporate locations) where attendance is not enough to justify offering the course. In fact, the Internet can be used to equalize the quality of education among schools in the same district with a minimal investment in technology, especially if the schools already have computers and Internet connections. All that is needed are teachers and administrators who are willing to experiment with new methods of instruction and information distribution.†

In your travels around the World Wide Web, you may have come across some amusing places where you can view a live shot of a famous landmark, someone's fishbowl or ant colony, or the lobby of an office building, a busy intersection, or some other public place. This is the simplest way to broadcast information. It involves the use of a stationary video camera plugged into your computer and so-called WebCam software such as EMULive from JcS Canada.

The possibilities for audio and video broadcasting—whether recorded or live—are limited only by your imagination. At the least, this method of content distribution can put new life into torpid Web pages; at its best, it can become an effective tool for delivering all kinds of information in a way that enhances the impact and retention rate of your message.

---

* VDOnet Corporation claims that Web pages that incorporate VDOLive content enjoy a surge of 10 times the visitors who stay 4 times as long.

† The idea of a camera in the classroom is anathema to most teachers. In fact, many public schools have strict prohibitions against any recording device being brought into the classroom, whether by a teacher or student. This idea is best implemented in private schools, colleges, and corporations where instructor performance is not such an emotional issue and where the motives for implementing this kind of technology are not as suspect.

System Requirements

In previous chapters, the system requirements for audio and video were discussed, including the roles of two key software components: the encoder that compresses the audio or video stream and the decoder that decompresses and plays the audio or video stream at the remote computer. Rather than repeat those discussions here, you can refer to Chaps. 2, 3, and 5.

It is worth noting, however, that the plug-in specification developed by Netscape Communications lets content developers define exactly how audio and video playback will work within a Web page. For example, you can have the audio or video start playing when the page is opened or embed the player into the body of the page rather than launching it as an external application. In the case of RealAudio, for example, you can even specify the appearance of the plug-in according to the controls you want displayed within your Web page. Through the use of the HTML <EMBED> tag, you can choose to display the Play and Stop buttons only, the control panel only, or all controls (Fig. 8.1). You can even specify the size of the control panel that will be displayed through the use of width and height values within the <EMBED> tag. A full description of how this is done can be found at the Web site of Progressive Networks at:

Figure 8.1
Netscape's plug-in specification allows the RealAudio Player from Progressive Networks to pop up within the browser when an audio link is selected, allowing the user to listen to audio content on demand and in real time instead of having to download the file and open it with an external player. Built-in controls even allow the user to pause, rewind, fast-forward, stop, and start the recording.

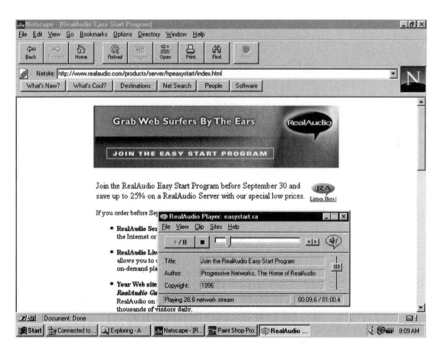

http://www.realaudio.com/products.ra2.0/features/plug-ins.examples.
html

Some plug-ins and helper applications, such as Xing Technology's
StreamWorks Player even provide information about the audio (or video)
stream (Fig. 8.2).

However, if you are interested in live broadcasting, a third component
might be required: *server software*, which delivers live data streams to remote users or provides advanced functionality to the player, or both. For
live audio broadcasting, server software is offered by such companies as
The DSP Group, Progressive Networks, VocalTec, and Xing Technology.
For live video broadcasting, server software is available from Apple Computer, VDOnet, and Xing. These products are installed on your server,
where listeners or viewers can connect to receive your broadcast. If you do
not have a server of your own, you might be able to lease space on the
server of your Internet service provider (ISP) or subscribe to a hosting service that specializes in Internet broadcasting. There are even firms that
will handle audio and video production for you to ensure the highest
quality presentation of your material.

However, you do not always need server-based software for audio or
video broadcasting. In fact, one of the first decisions you will have to

Figure 8.2
With Xing Technol-
ogy's StreamWorks,
configured either as a
browser plug-in or
helper application,
users can call up in-
formation about the
audio or video stream
they are playing.

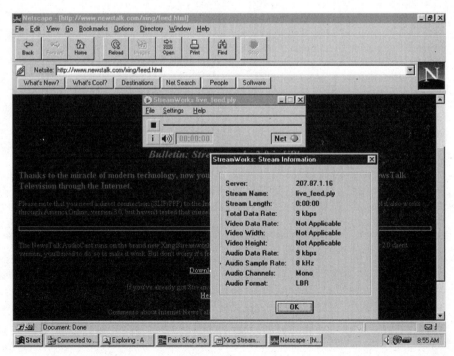

make is choosing between server-based and serverless broadcasting. The choice will determine both the number of simultaneous users a system can service and the features users will be able to obtain. If you expect to distribute no more than five simultaneous audio streams, for example, you can get by without audio or video server software. This is a pretty safe assumption, especially if you are a lone entrepreneur just getting started using this medium. You can always graduate to a server-based solution later, if demand warrants. Since most vendors of this kind of software offer both serverless and server-based broadcast products, you can stick with the same vendor as you grow.

On the other hand, if you are an experienced developer who has a continuous supply of quality material that seems to be in hot demand—sports programs and rock concerts, for example—you may need to support dozens of simultaneous audio/video data streams. In this case, you can safely assume that server software is required. Be aware, however, that server software typically represents the most significant system cost.[*] Xing Technology's StreamWorks Transmitter, for example, is a dedicated hardware/software system that costs $6500, enabling you to deliver live or on-demand audio/video programs across any IP-based network (Internet, intranet, WAN, or LAN).

Control of end-user features is another reason to go with a server-based system, especially when using long audio/video files. Server software is required to support the player's ability to scan or fast-forward recorded programs, something users will find extremely convenient if your files exceed a few minutes in length. VocalTec's server software provides this feature, as well as Progressive Networks and Xing. With Xing's StreamWorks Transmitter, for example, you can even control such video parameters as picture size, sampling rate, data rate, and sophisticated MPEG variables such as frame intervals.

Nonserver Audio Solutions

A nonserver audio broadcast solution, such as VocalTec's IWave and The DSP Group's TrueSpeech, has several advantages. The big advantage is that

---

[*] Another significant cost may be for the license to broadcast live performances or recorded material that is owned by someone else. The topic of copyrights and copyright infringement is beyond the scope of this book. It is best to seek the services of experienced legal counsel for assistance on these matters.

these products do not require extra-cost server software for sending audio data streams. Unlike other audio-on-demand applications, which use the User Datagram Protocol (UDP) to send the audio data stream, IWave and TrueSpeech are TCP/IP-based audio-on-demand applications. This is what allows them to use any available Web server as their data source, thus eliminating the cost of purchasing, installing, and maintaining a dedicated server.

Since these products can use any Web server, with all its existing protocols, even users located behind a corporate firewall can use a Web browser to access IWave- and TrueSpeech-based audio content. No reconfiguration of the firewall is necessary. In contrast, other audio-on-demand products that use UDP will require a reconfiguration of a corporate firewall or the installation of a proxy server to allow access to the audio stream on the Internet. Given corporate concerns about security and employee productivity, these measures are becoming increasingly necessary.

Although audio-on-demand products usually do not require dedicated server software, they may require that certain pieces of code be loaded at the server. This can be your corporate Web server or the server of your Internet service provider (ISP). For example, by installing a simple common gateway interface (CGI) file supplied with VocalTec's IWave package, listeners of your audio content can request random access to any segment of the audio file and give the user the ability to skip uninteresting parts of the music or speech by using the rewind or fast-forward controls, reducing unnecessary data transfers.

VocalTec's Internet Wave

VocalTec's IWave has two main parts: a server utility that includes an encoder that works in conjunction with standard Web servers and the IWave player which can be registered with any Web browser, including Netscape and Spyglass, as a helper application. Both components are available free of charge at the company's Web site at:

http://www.vocaltec.com

VocalTec also offers an encoder for live audio source compression, but charges for it.

The encoder compresses data from Windows WAV and UNIX AU files using a compression algorithm that is similar to the one developed by VocalTec for the Internet Phone discussed in Chap. 2. The compressed data is then stored on a Web server for retrieval. Once installed, the IWave helper application automatically recognizes and plays back audio stored at Web sites.

According to VocalTec, those with 14.4 Kbps modems will experience audio with quality similar to that of AM radio broadcasts, while those with 28.8 Kbps modems will hear audio of almost-FM quality. Users who call into Web sites with IWave players have the ability to choose what sections of the audio clips they would like to hear.

Since IWave acts like any other Web application, audience size for an IWave broadcast is limited only by a server's connection bandwidth. In addition, you can set passwords to limit access to IWave sites or audio files. This capability is useful for developing electronic commerce applications that involve the sale of recordings, for example.

You use the IWave encoder software to compress standard audio files into the Internet Wave audio compression format and then use the server utility to add the Internet Wave programming to your Web site and pages. When this utility is installed on your server, it gives other users who play the audio content full control over playback, and lets them fast-forward or rewind.

You can record speech and music through a microphone as well as from an external source, such as a CD player, tape recorder, DAT, or any other external audio source. The WAV file can be edited to remove unnecessary content. This is done by opening the audio file in an editing application such as Cool Edit or GoldWave. These tools also let you add special effects. Once the file has been edited to your satisfaction, all you need to do is use the IWave encoder to format and compress the recorded WAV file.

The choice of compression type should be made according to the Internet connection speed available and the sampling rate of the input WAV file. Within the IWave framework, you can choose from among several recording formats that will determine the quality of your audio content. You can offer both standard and high-quality audio on your Web page, letting each user select the best audio format which he or she is capable of hearing. If you are going to use only one format on your page, you should select the format which requires the least amount of bandwidth while still producing satisfactory sound quality. The following IWave compression options are available:

VSC77. Offers the lowest recording quality, which is suitable for Internet connection speeds as low as 9.6 Kbps. Compression is applied to files recorded at a 5.5-kHz sampling rate (recommended) and up. This format provides acceptable speech recording quality and can be used by all users due to its low bandwidth requirements.

VSC112. Offers average quality recording, which is suitable for Internet connection speeds of 14.4 Kbps and up. Compression is applied to files recorded at an 8-kHz sampling rate (recommended) and up. This format provides average speech recording quality and reasonable music quality and can be used by almost all users due to its low bandwidth requirements.

VSC154. Offers good quality recording, which is suitable for Internet connection speeds of 28.8 Kbps and up, while still leaving a substantial amount of bandwidth available. Compression is applied to files recorded at an 11 kHz sampling rate (recommended) and up. This format provides good speech recording quality and good music quality. It can be used only by those with 28.8 Kbps modems or higher-bandwidth connections due to the higher bandwidth requirements, but it will still leave some of the bandwidth for other applications to use at the same time.

VSC224. Offers high-quality recording, which is suitable for Internet connection speeds of 28.8 Kbps and up. Compression is applied to files recorded at a 16-kHz sampling rate (recommended) and up. This format gives an excellent speech recording quality and excellent music quality and can be used only by users with 28.8 Kbps modems or higher-bandwidth connections due to the higher bandwidth requirements.

Although Internet Wave does not require a special server, you do need to have a system administrator or Internet service provider make a few simple changes in the server so that it can recognize the resulting Vocal-Tec file types, which have VMD and VMF extensions. To have these file types recognized by an NCSA HTTPD server, the following lines must be added to the conf/srm.conf file:

```
AddType application/vocaltec-media-desc.vmd
AddType application/vocaltec-media-file.vmf
```

Alternatively, these lines can be added to the conf/mime-types file without the AddType prefix.

On a CERN HTTPD server, the following lines must be added to the /config/httpd.conf file:

```
AddType .vmd application/vocaltec-media-desc 8 bit
AddType .vmf application/vocaltec-media-file binary
```

The next time the server is restarted or the configuration file is reloaded, your audience will be able to play the Internet Wave audio files you have created.

The DSP Group's TrueSpeech

Another product that does not require a dedicated server for on-demand audio is The DSP Group's TrueSpeech. According to the company, the steps to create your own audio files are as follows:

1. *Create the WAV file.* You can create a PCM-encoded WAV file by using the Sound Recorder application that comes with Windows. (In Windows 95, the Sound Recorder is located at Start> Programs> Accessories> Multimedia> Sound Recorder.) Although Sound Recorder supports other encoding schemes, the TrueSpeech compression algorithm has been optimized for a sampling frequency of 8 kHz with 16-bit resolution. Any multimedia board which is a Sound-Blaster 16—compatible will support this format. If your audio equipment does not support this format, you can use audio editing tools such as Cool Edit or Gold Wave to make the necessary conversions.

2. *Convert the file to a TrueSpeech-encoded WAV file.* To convert the PCM-encoded WAV file to a TrueSpeech-encoded WAV file, you can use the Sound Recorder included with Windows 95 or NT, since Microsoft supports TrueSpeech there. If TrueSpeech does not appear as an option, you can download a converter from The DSP Group's Web site at:

> www.dspg.com/allplyr.htm

In Sound Recorder, simply open the PCM-encoded WAV file created in step 1, select the TrueSpeech format, and then use the Save-As command to create a new filename with a WAV extension. Then change the file type to DSP Group TrueSpeech (Fig. 8.3), and then save.

The new WAV file will be encoded in the TrueSpeech audio format which is compressed by a factor of 15 from the original WAV file. For those who are using Windows 3.11, a PCM-to-TrueSpeech conversion utility can be downloaded from The DSP Group's Web page at:

> http://www.truespeech.com

This conversion utility will accept PCM-encoded WAV files only if sampled at 8-kHz/16-bit PCM format.

You can link the TrueSpeech WAV file directly into your Web page and it will play streaming audio with its built-in capabilities. Microsoft Internet Explorer 3.0 browsers can also begin streaming audio in the background when a page is loaded. To achieve this, just insert the following sample tag at the top of your Web page:

Figure 8.3
PCM-encoded WAV
files created within
Microsoft's Sound
Recorder can be
turned into com-
pressed (i.e., 15:1)
WAV files in The DSP
Group TrueSpeech
format.

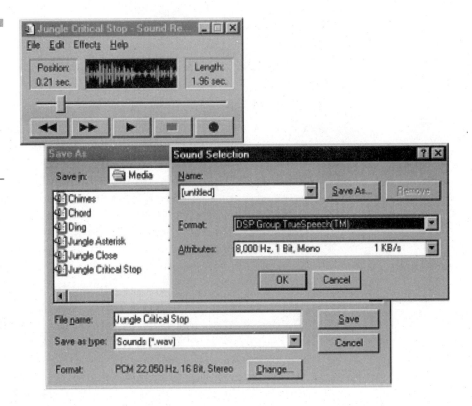

<BGSOUND src="filename.wav" loop="Infinite">

3. *Create an associated text file.* A one-line text file must be created which is associated with the TrueSpeech-encoded WAV file. You can use a standard text editor such as Notepad and save this file with the TSP extension. This is done so that the browser can launch the True-Speech Player when a TrueSpeech-encoded WAV file is about to be sent. The TSP file associated with a TrueSpeech-encoded WAV file should contain the following HTML line, which is case sensitive:

TSIP>>URL/filename.wav

The preceding line should not include the HTTP:// characters that are usually found in a URL (Uniform Resource Locator). For example, www.ddx.com would be correct and HTTP://www.ddx.com would not be correct. For example, let's say that you have created a PCM-encoded WAV file and have converted it to a TrueSpeech-encoded WAV file. This file might be named sample.wav and it might be located in a subdirectory called audio at the server. The one-line TSP text file would contain the following line:

TSIP>>www.ddx.com/audio/sample.wav

You can name the file containing this line *sample.tsp.*

4. *Link the text file into your Web page.* The TSP text file should now be referenced within your Web page and associated with any image or text you specify as a hypertext link. When a visitor clicks on the sample.tsp link, the TrueSpeech Player will start. Upon spawning the player, the TrueSpeech Player reads the location of the TrueSpeech sample.wav file from the sample.tsp file. The Player then accesses and plays this file as it is being downloaded so the user can listen in real time.

5. *Configure the server.* Because the TSP file extension will probably not be recognized by HTTP servers running on UNIX or Windows (unless someone has already done the necessary mapping to support TrueSpeech audio files), the server must be configured for MIME-type application/dsptype. To do this on UNIX servers, only one line needs to be introduced in the HTTP server configuration file mime.type as follows:

```
application/dsptype tsp
```

On a CERN HTTP server, the configuration line should read:

```
AddType.tsp application/dsptype binary 1.0
```

If your Web page is located on a remote HTTP server, you must contact your Internet service provider and have it add these lines. If your Web page runs off a corporate server, you must contact your system administrator to have this done.

Server-Based Audio Solutions

To play real-time audio, some products such as Progressive Networks' RealAudio require server software in addition to an encoder and player. The *encoder* enables content developers to put standard PC and UNIX audio files into the RealAudio format, while the *player* allows you to listen to audio files that have been encoded in the RealAudio format. The server enables audio streams to be sent over the Internet to RealAudio players.

Although you can directly reference a RealAudio file in your Web page, activation of the hypertext link merely initiates a download of the file to your computer's hard disk before it can be played. To enable users to play audio files in real time via the RealAudio plug-in discussed earlier requires the construction of metafiles that contain the URL of the Real-Audio file. The *metafile* serves as the connection between the Web server and the RealAudio server software. Unlike the usual HTML hypertext link, the metafile does not display information through the browser. Instead, it provides the audio file's URL to the RealAudio player.

If you are committed to using the RealAudio format, a very useful reference to obtain is *The RealAudio Content Creation Guide* published by Progressive Networks. This 46-page guide is available in Adobe Acrobat format from the company's Web site at:

http://www.realaudio.com/help/server/index.html

Just download the file, print it out, and put it in a three-ring binder. The guide includes detailed information, illustrations, and plenty of HTML examples that step you through encoding audio files, configuring the Web site, producing live events, and creating synchronized multimedia presentations. These presentations can be as simple as a narrated slide show of your Web page or as complex as a multiframe training program that the viewer can control.

The bandwidth negotiation feature of the RealAudio server allows RealAudio players to receive optimal sound quality based on their connection speed. A user who connects to a RealAudio server at 14.4 Kbps will receive RealAudio encoded with the 14.4 algorithm, while a user who connects to a RealAudio server at 28.8 Kbps will receive RealAudio encoded with the 28.8 algorithm. The RealAudio server automatically detects the connection speed of the RealAudio Player and chooses which RealAudio file should be sent.

Only one generic link is required to a particular content file on your Web site. Without bandwidth negotiation, to provide content in both 14.4 and 28.8 formats, your Web site would require two hypertext links and two metafiles—one to a 14.4 file and one to a 28.8 file. Using bandwidth negotiation, however, you only need one hypertext link and one metafile. This not only makes Web page creation easier, it allows you to keep your audio content better organized.

Nonserver Video Solutions

Apple offers a nonserver solution through its QuickTime Conferencing software, which allows your Macintosh to become an active broadcast station to a handful of viewers. You can broadcast live or recorded events over the Internet through your Macintosh* configured with MacTCP or TCP/IP. Viewers access your broadcast using QuickTime TV, a free trial version of which is available for download at:

http://qtc.quicktime.apple.com.

For best results, however, QuickTime TV users need an ISDN connection to the Internet that runs at 112 Kbps (provided by two 56-Kbps bearer channels) or better. And if you want to broadcast to more than just a handful of viewers, you will need to have a server component called a *reflector* (discussed later).

To broadcast a real-time stream to multiple viewers, your computer must deliver each packet of digitized audio and video to all destinations. Traditional Internet protocols use *unicast packet delivery;* that is, each packet that leaves your computer has an explicit host address. To deliver your broadcast to five viewers using unicast, your machine must send five copies of the program, packet by packet (Fig. 8.4). As more and more viewers join in, your computer's network connection becomes increasingly clogged by this outbound traffic. The computer's CPU becomes increasingly burdened with copying each packet to the network to serve all the viewers.

Newer multicast protocols are supposed to offer a better way of handling packets to minimize network traffic. In this scenario, routers on the Internet are configured for multicasting in order to perform the replication of packets, but only when necessary. For example, if you are broadcasting from a location in Atlanta to multiple destinations in and around Los Angeles, a single stream of digitized audio and video would reach an Internet router in Los Angeles and be duplicated there to reach the various destinations on adjacent subnets. Instead of duplicating packets in

---

* You should use an 8500 system as the broadcast station. Although an 8100 will do, your minimum audio rate may be 22 kHz, whereas the minimum audio rate on an 8500 is 11 kHz. You can compress voice as much as 8:1 with the G.728 audio compression standard, but this is very CPU-intensive. While it may sound good at extremely low bit rates, it will choke any source machine slower than an 8500. The bottom line is that the 8500 will give you more flexibility in developing the audio component of your broadcast.

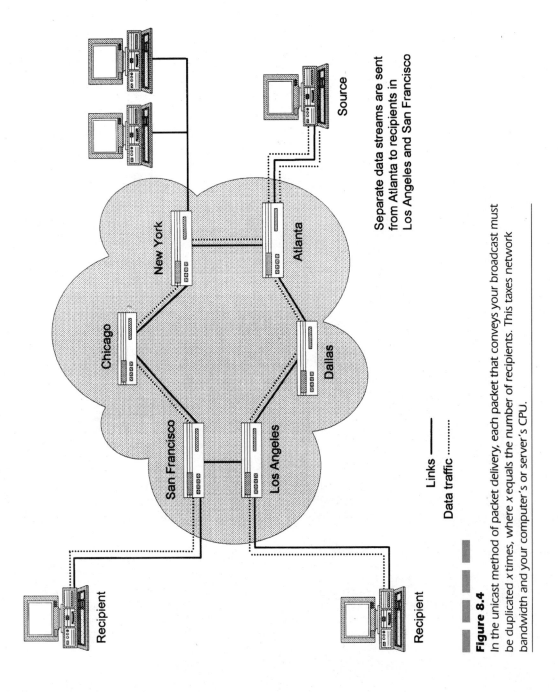

Separate data streams are sent
from Atlanta to recipients in
Los Angeles and San Francisco

Links ——————
Data traffic ·············

Figure 8.4

In the unicast method of packet delivery, each packet that conveys your broadcast must
be duplicated *x* times, where *x* equals the number of recipients. This taxes network
bandwidth and your computer's or server's CPU.

Atlanta, they are duplicated at the very last router (Fig. 8.5). This conserves bandwidth on the Internet and eases the processing burden of the server.

Multicasting is still not widely implemented on the Internet, with the notable exception of an overlay network known as the MBone. It is not only difficult and time-consuming to configure routers for multicasting, but the possibility of a lot of real-time traffic burdening their routers makes system administrators balk at the idea of supporting multicast protocols.

The next best thing to multicast is to use a *reflector*, which acts as a proxy for broadcasters. Apple offers a reflector that works in conjunction with its QuickTime Conferencing software. Instead of calling your Macintosh, viewers connect to the QTC Internet Reflector,* which takes care of replicating the audio and video streams given to it by your Macintosh. Each reflector can support multiple viewers—the exact number of viewers it can support depends on the rate of the data stream, the speed of the host machine, the type of network interface it has, and the network bandwidth allocated to the reflector by your Internet service provider or corporate system administrator.

You can determine the number of viewers you can support by dividing your allocated bandwidth by the estimated bandwidth of the stream(s) you will be sending—just subtract one to account for the source stream from your computer. For example, if you have been given a T1 line which provides 1.544 Mbps (or 1500 Kbps) of total bandwidth and you plan to send a 150-Kbps signal, you can support nine viewers, according to the following formula:

$$[1500 \text{ Kbps} \div 150 \text{ Kbps}] - 1 = 9 \text{ users.}$$

To serve hundreds or thousands of viewers, you will want to distribute the network load by installing multiple reflectors at geographically disbursed locations. To do this you will need login accounts on UNIX machines to serve as reflectors and compiled versions of the reflector software for each type of machine (e.g., SunOS, Solaris, FreeBSD, NextStep, Irix, A/UX, AIX). Detailed instructions for setting up your own broadcast network and configuring the reflector can be found at Apple's QuickTime Web site at:

---

* The QTC Internet Reflector is not a meeting point like the Enhanced CU-SeeMe reflector described in Chap. 3. Rather, it functions as a splitter, with a single broadcast source which the QTC daemon routes to multiple viewers. The reflector connects to the broadcaster's Macintosh, receives the audio and video streams from that machine, and performs the replication necessary to deliver the stream to multiple recipients.

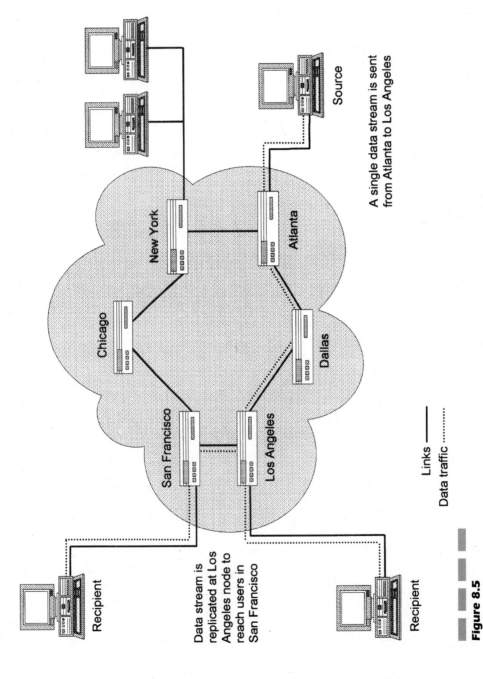

Source

A single data stream is sent
from Atlanta to Los Angeles

New York

Atlanta

Chicago

Dallas

San Francisco

Los Angeles

Recipient

Data stream is
replicated at Los
Angeles node to
reach users in
San Francisco

Recipient

Links ———
Data traffic ··········

Figure 8.5

In the multicast method of packet delivery, each packet that conveys your broadcast is
duplicated at the most appropriate point in the network. This conserves network bandwidth and relieves the processing burden of the server.

http://devtools.apple.com/qtc/qtcddocs.html

When your broadcast network is set up, each reflector receives its feed from the nearest reflector. Since the Internet is being used for the broadcast, reflectors can be located at servers anywhere in the world. While distance is not a factor, your wallet may be. After all, you will have to pay for server space and bandwidth at every site where you want to install a reflector.

Server-Based Video Solutions

A live video broadcast almost always requires a server-based solution, especially when a significant number of viewers is expected. There are two ways to cover a live event from a remote site: you can send the analog signal from the remote site to the transmitter at the local site or you can put the transmitter at the remote site and establish a network connection to the local site. Xing Technology's StreamWorks Transmitter software allows you to do both for video as well as audio broadcasts.

Xing Technology's StreamWorks

There are two types of StreamWorks Transmitters available: Audio Transmitter (ATRANS) and Audio-Video Transmitter (AVTRANS). The ATRANS system accepts analog audio input and performs real-time audio encoding and transmission. The AVTRANS system includes all the functionality of ATRANS, plus the ability to perform real-time MPEG video encoding and transmission. When the StreamWorks Transmitter is added to a Stream-Works Server, the server becomes a live broadcast center capable of delivering live (or on-demand) video and audio to a large audience using the StreamWorks Player.

Complementing StreamWorks Transmitter is XingMPEG Encoder, a software solution for creating MPEG audio and video files for on-demand delivery over the network or viewing from local disk. XingMPEG Encoder converts the digital input from popular capture and editing tools, such as motion-JPEG capture cards and nonlinear editing systems, into MPEG-compliant audio and video streams for use on the StreamWorks Server.

StreamWorks Server can support users connected at any data rate, including 14.4 and 28.8 Kbps over ordinary phone lines and higher-speed ISDN connections. The StreamWorks Player provides the high-quality audio and video and is capable of up to full-screen, full-color, full-motion video with CD-quality, 44-kHz audio.

The server delivers audio/video streams to players over the Internet in the form of UDP packets. In a unicast transmission, the stream's destination is the IP address of a receiving node on the network. The server must send a separate stream to every recipient. In a multicast transmission, the stream's destination is an IP-multicast address. This is usually the address of a router which replicates the data stream so it can be passed on to recipients on adjacent subnets. As noted earlier, this minimizes bandwidth consumption and distributes the processing burden of replicating the data stream. Unicast transmission is used when you want to broadcast to a small group of known recipients. Multicast is used when you want to broadcast to a wider audience. However, unicast transmission is required in any network configuration that does not support multicast traffic.

The StreamWorks Server has the ability to reduce the data rate of streams to meet the bandwidth requirements of remote StreamWorks Players. This allows you to have a single source—an on-demand file or a live transmission from a StreamWorks Transmitter—that is served at various data rates, according to the connection speed of the various StreamWorks Players.

A capability called *stream propagation* allows your StreamWorks Server to serve audio/video streams originating on another StreamWorks Server and to have the stream propagated by other StreamWorks Servers. Propagation maximizes the size of your potential audience when you coordinate with other StreamWorks Server sites. For example, if your StreamWorks Server's maximum bandwidth is limited to 1.544 Mbps over a T1 line, and you want to serve a stream with a data rate of 10 Kbps, your potential audience is only 154 people. But if you arrange for the propagation of your stream by another StreamWorks Server that has a maximum bandwidth of 45 Mbps (the bandwidth of a typical T3 line, which is equivalent to 28 T1 lines), your potential audience increases to over 4500 people.

Stream propagation is also an effective way to diversify your Stream-Works Server's available content. By propagating streams from other StreamWorks Servers, you can offer your audience a greater selection of available streams.

Broadcasts originating from StreamWorks Servers may be accessed through firewalls. A *firewall* is software and/or hardware that prevents direct communications between a corporate LAN and the greater Internet for security purposes. However, direct communications between the LAN and the Internet can be permitted on a selective basis by assigning a port in the firewall or running a proxy on the firewall. StreamWorks streams can be accessed through firewalls via either method.

In the first method, the network administrator can configure the firewall for incoming and outgoing UDP traffic on port 1558, which is Xing's InterNIC-registered port. Since only StreamWorks traffic has access to this port, security will not be breached. In the second method, the network administrator can obtain the source code for an appropriate proxy from Xing Technology by sending e-mail to:

streams@xingtech.com

A *proxy* is server software that gives users cached copies of Internet material; in this case, broadcasts from StreamWorks Servers. In addition to improving performance, the proxy can be used to limit access to the broadcasts.

VDOnet's VDOLive

VDOnet offers a server-based solution for on-demand broadcasting of video content by using VDOLive products as follows:

- Capture and compress your video and audio data using VDOLive Tools
- Store the resulting files on the VDOLive Server computer
- Create an HTML link from your Web page to the video file

When viewers access your Web page, they can click on the video links you have provided to see full-color motion video in real time via the VDOLive Video Player.

The VDOLive Video Server can deliver video files to multiple users at the same time and adjust the quality of video according to the bandwidth available to each user. It continues to scale up or down as long as the user is connected. This dynamic scalability is built into the compressed file and provides the best possible video quality at any particular moment. Data about each user connection to the VDOLive Video Server is recorded in a log file. This data can be used for marketing, billing, and other business purposes.

You can only install the VDOLive Video Server if you have super-user privileges on a UNIX computer. If you do not have such privileges, you must have your system administrator install the server software for you or, if you have a login account with an Internet service provider, have the Web site administrator install the software on that server. If the Internet service provider does not have the resources to accommodate your broadcasting needs, you will have to look for a hosting service that specializes in supporting broadcasters.

Linking to a video from a Web page will launch the VDOLive Player when a user clicks on the link. In this case, the VDOLive Player acts as a helper application. This method will work with all Web browsers that can use helper applications (e.g., all versions of Netscape, Mosaic, Internet Explorer, etc.). The link uses the normal HTML anchor tag and should point to a file with the VDO extension, as in the following example:

<A HREF="celebrity_talk.vdo">Click here to see the video</A>

The VDO file contains one line pointing to the location of the actual video clip which has the AVI extension.

Embedding a video in an HTML page is done using the EMBED tag. The tag requires several arguments:

- SRC="name of VDO file pointing to the video"
- HEIGHT=plug-in height in pixels
- WIDTH=plug-in width in pixels
- STRETCH=TRUE/FALSE (fits plug-in content to specified size of plug-in window)
- AUTOSTART=TRUE/FALSE (video starts playing immediately or waits for mouse-click)

An entire EMBED tag might look like this:

<EMBED SRC="celebrity_talk.vdo" WIDTH=160 HEIGHT=
    128 STRETCH=TRUE AUTOSTART=TRUE>

VDO files contain the URL for the video file. The VDOLive Player uses files in AVI format. The URL contains the following information:

- *Resource type.* The resource type for VDOLive videos is vdo://.
- *Server address.* Name or IP address of the server where the AVI file is located.
- *Port number.* The TCP port used by the server (default is 7000).
- *Full path to AVI file.*

The URL syntax is:

vdo://server_address:port/path

An actual URL might be written as:

vdo://vdo1.vdo.net:7000/pub/movies/celebrity_talk.avi

You can create VDO files using any text editor or VDOnet's online VDO file generator to automatically create one.

If you need help deciding whether to commit to using VDOLive, a very useful reference to obtain is *VDOLive Video Server and Tools Manual* published by VDOnet Corporation. This 150-page guide is available for download in Adobe Acrobat format from the company's Web site at:

http://www.vdo.net/tech/

VDOnet also offers the VDOLive Personal Server, configurable for delivery of two simultaneous video streams. The software handles two-minute video clips and serves up to 28.8 Kbps modem connections. Designed for the communications needs of individuals and small organizations, the product is available for free downloading from VDOnet's Web site.

Online Resources

The following table provides the Web links of the major Internet video-conference software vendors from which you can download working or evaluation copies of their software. The Web pages also contain such information as platforms supported, system requirements, product features, installation instructions, and troubleshooting advice. Since the technol-

Developer	Product	Web page or FTP site
Apple Computer	QuickTime TV	http://qtc.quicktime.apple.com/
Destiny Software Productions	Radio Destiny Broadcast LIVE!	http://web20.mindlink.net/destiny/live.htm
JcS Canada	EMULive	http://www.jcs-canada.com
Progressive Networks	RealAudio	http://www.realaudio.com
The DSP Group	TrueSpeech	http://www.truespeech.com
VDOnet Corp.	VDOLive	http://www.vdo.net
VocalTec	Internet Wave	http://www.vocaltec.com
Xing Technology	StreamWorks	http://www.xingtech.com

NOTE: This information, as well as updates, can be found at http://www.ddx.com/mgh.shtml.

ogy is moving rapidly and new products are becoming available all the time, you may want to browse these Web pages periodically for the latest developments.

Conclusion

Broadcasting involves a commitment of time and effort to develop quality content for on-demand or live audio/video distribution. Depending on the hardware and software you select, broadcasting can also entail significant initial and ongoing costs which should not be taken lightly. Even after you've gone to all this trouble, you may not be able to sustain an ongoing audience that appreciates the content you have to offer, let alone make any money from it. What little money is being made today from Internet broadcasting seems to be from the so-called adult entertainment services.

That market aside, the largest audiences for live rock concerts broadcast over the Internet rarely exceed 1000 listeners or viewers at a time. So do not take too seriously vendor claims that you will be able to "reach millions" of Internet users with your material. Despite being available for several years over the MBONE, the whole area of broadcasting over the Internet is still in its developmental stage. All this may change in the near future, however, as more bandwidth is added to the Internet, as compression technologies continue to improve, and as big companies in the traditional media start seeing the potential of Internet broadcasting to augment their existing print and broadcast offerings.

Nevertheless, there are some things you can do to increase your chances of success in Internet broadcasting:

■ Add links to relevant useful information to your Web page, including the dates and times of your broadcasts, system requirements for listening or watching, where to get the software, how to configure browsers, and how to connect to the broadcast.

■ When your Web page is ready, publicize the event by filling out the forms of major search engines such as Alta Vista, Excite, Lycos, and Yahoo so they will scan your Web page and feed its content into their search databases.

■ Send e-mail to relevant mailing lists and post announcements to appropriate newsgroups.

■ You can send press releases to the various electronic magazines on the Web, as well as to traditional media. Include the URL of your Web page and your e-mail address on all announcements, electronic or print.

Taking these steps will provide the exposure you need to attract an audience. Keeping that audience will depend on the quality, timeliness, and originality of your content.

Collaboration Over the Internet

Introduction

The global nature of the Internet makes it particularly attractive for keeping in touch with friends, colleagues, business contacts, and even strangers who share your interests. Other chapters discuss how to do this very economically with various chat, e-mail, audio- and videoconferencing, and news reader products. However, with collaborative tools that operate over the Internet, several people at far-flung locations can work together interactively, exchanging information and sharing ideas, for the purpose of achieving a common objective. The collaborative process can even draw upon existing applications and data which can be viewed and annotated by everyone.

Collaboration software can be used for such purposes as brainstorming new ideas, devising schedules and plans, hammering out design issues, sketching out customer solutions, revising documents for publication, and troubleshooting problems on remote PCs. In the process, participants can mark up shared documents with highlights, sticky notes, and other tools; pass files and messages back and forth; view each other's screens; bring up supporting documents and images for viewing within the shared workspace; and send e-mail. Some collaborative products even offer shared Web surfing, audio communications, videoconferencing, and security features. All of this activity occurs over a common Internet access link. For better performance, some collaborative products support direct connections over ordinary dial-up lines via modems or ISDN connections via terminal adapters, in which case, per-minute usage charges may apply.

Types of Products

Collaboration products fall into several main categories. For purposes of discussion, these categories include products for remote access, whiteboards, discussion tools, application sharing, and calendaring.* Some industry

---

* There is actually another category of collaborative product: the workflow application. *Workflow* refers to a standardized sequence of tasks which occurs within a company and which has been established to improve productivity. For instance, one type of workflow application is a *help desk*. When a company receives a technical support call from an employee or customer, it usually has a standard procedure for entering, tracking, and resolving the problem. Workflow software automates these processes. This topic is beyond the scope of this book because workflow products are specialized, expensive (as much as

experts might consider audio- and videoconferencing software and chat programs as collaboration products. There's no argument here, except that these products are so feature-rich that they deserved chapters of their own.

Some collaboration products overlap several categories. Thus you might find a whiteboard product that includes audio, chat, and group discussion tools, or a videoconferencing product that includes a whiteboard capability. Other products, such as VocalTec's Internet Conference, use a whiteboard as the main facility from which all connections and other activities—including chat, file transfer, and application sharing—are launched.

There really are no firm rules about how collaboration products are categorized. Some products, such as NetManage's ChameleonNFS/X, defy easy categorization because they are actually a suite of applications that provide host connectivity, e-mail and messaging, file and printer sharing, Internet access, workgroup collaboration, and desktop management.

With remote-access products, the files and directories of a remote computer can be viewed and copied, the remote PC's screen can be observed, and the remote PC can be controlled by the local PC—all over the Internet. Among the uses for this kind of software are remote training, software demonstration/evaluation, remote support, and remote access to corporate databases.

A whiteboard provides a shared canvas in which drawings, documents, and image files can be viewed by participants at remote locations and annotated using various drawing tools. The annotations appear on all participants' screens. You can also copy material from files in other programs and paste them into the canvas for review and markup. Most programs keep pasted data separate from annotations, so you can mark up the pasted version and then clear the annotations without losing the original document. You can also print and save the whiteboard file for future reference. Advanced whiteboard features include screen capture capabilities; a remote pointer that is used to highlight portions of the document without actually marking it; screen synchronization, which lets everyone look at the same location within a file; and image compression, which speeds data transfer between participants.

Application sharing lets multiple users view and edit a file in its original form, using that application's tools. In application sharing, you might

---

$40,000), and intended for corporate use over private intranets. Companies that offer workflow products include Action Technologies (http://www.actiontech.com), Lotus Development Corporation (http://www.lotus.com), Ultimus (http://www.ultimus1.com), and ViewStar Corporation (http://www.viewstar.com).

dial up a host computer that is running a program such as Excel, Word, or PowerPoint for the purpose of reviewing and revising a file.

Whiteboard and application sharing products may include file transfer and text chat capabilities to enhance the collaborative process and some—such as Microsoft's NetMeeting—may even include an audioconferencing capability as well. Some products that originated as text chat programs—such as Tribal Voices' PowWow—now include whiteboard and audioconferencing capabilities (see Chap. 2). And then there are companies such as VocalTec which started out with Internet telephony products that now offer full-fledged Internet collaboration products.

At least one company, InSoft, started with videoconferencing and document conferencing software for corporate networks and migrated its data collaboration technology to the Internet. As part of its Interactive Internet Collaboration Environment (IICE), InSoft offers CoolTalk, a shared whiteboard application and audioconferencing tool that allows users to simultaneously view and mark up documents or images while conducting a voice conversation. Another component of IICE is CoolView, which adds a five-frames-per-second videoconferencing feature to the voice and data sharing capabilities in CoolTalk.

There are two ways in which collaborative applications operate over the Internet. Some work in real time; that is, participants share information and get screen updates virtually instantaneously. Others use the store-and-forward method of information transfer, meaning that a file must be downloaded before participants can view changes made to the document. InSoft's CoolTalk is an example of the former, while Notes from Lotus Development Corporation, is an example of the latter. If spontaneity is important in your collaborative efforts, then products that offer real-time information transfer and screen updates will suit your needs better than those that use the store-and-forward approach. However, if you will be working with big files for a long period of time, it might actually speed the collaboration process if each participant had a copy of the document saved to disk. That way, there will be minimal delay when participants have to go from one page to another and back again.

System Requirements

In general, there are no special system requirements for running collaborative applications over the Internet that have not already been discussed in previous chapters. Basically, all you need is an Internet access account and a modem-equipped computer. However, if you want to take advantage of the audioconferencing capabilities of some collaborative applications, you will need a sound card, microphone, and speakers or headset, as discussed in Chap. 2. And if you want to use collaborative applications that offer videoconferencing, you will have to add a video camera, as discussed in Chap. 3.

Of course, you can get peak performance from collaborative applications if you have an ISDN access connection to the Internet. ISDN basic rate service is delivered over a digital line and provides two 64-Kbps channels that can be used separately—one for the Internet connection and the other for your phone—or together to form a single higher-speed connection of 128 Kbps. Regardless of how you apportion the channels, you will require a terminal adapter that connects your PC to the digital line.

If you're about to buy a new modem and want to leave the option open for ISDN in the future, you should consider buying a device that combines a 33.6-Kbps modem with an ISDN terminal adapter in a single unit. U.S. Robotics, for example, offers such a product for less than $400. If you already have a good modem and ISDN is available in your area, you can buy an ISDN terminal adapter for $300 to $400. Some are available for as little as $200, but they do not support key ISDN features such as distinctive ringing for multiple-line appearances. Cardinal Technologies is among the vendors of this type of bare-bones terminal adapter.

Remote Access Over the Internet

The simplest collaborative applications are those that allow you to access a remote computer over the Internet for real-time screen sharing and file transfers. You can use this type of software to help resolve problems at remote locations, train other computer users, and to collaborate on projects by viewing documents on each other's screens.

Farallon's Timbukto Pro

You can try real-time screen sharing and file transfers over the Internet by downloading two free applets from the Web site of Farallon Computing at:

http://www.farallon.com

There, you can find Look@Me and FlashNote, a pair of applets designed to let users remotely connect to and access other computers over the Internet. Look@Me is a simple screen-sharing tool that lets two users view the same file, application, or document at the same time on their PCs. Annotations and edits made on the host PC can be viewed in real time at the remote PC.

FlashNote enables one user to send a file with an annotated text message to another user over the Internet. Both applets are based on the company's Timbuktu Pro remote-control software, which enables Windows and Macintosh users to collaborate on projects, conduct interactive conferencing, and send files and annotated text in real time over the Internet.

If you like the way the free applets perform, you can purchase Timbuktu Pro which includes advanced security features, both personal and shared address books that allow you to store frequently used connections, and the capability to remotely control another computer with your computer's mouse and keyboard.

Transferring files across the Internet is normally very cumbersome, involving user accounts, passwords, compression, encoding, and other processes. Timbuktu Pro eliminates all this by enabling users to establish more efficient peer-to-peer connections over the Internet. Users establish a Timbuktu Pro connection to another individual simply by specifying an IP address or IP name,* selecting the file to transfer, and using the Send or Exchange command to initiate the process.

File and directory transfers made with the Send command are accompanied by a message called a *FlashNote.* This is a pop-up icon that appears on recipients' desktops. Double-clicking on the icon maximizes the Flash-Note window which shows a listing of received files and/or directories. Here, you can open the files for viewing, move them to another directory, delete them, or send a reply.

With the Exchange command, file transfers can be bidirectional and one user's access to another user's directories and files can be virtually unlimited. Exchange allows you to copy files to any directory or from any directory on the remote disk. You can also create directories on and remove files or directories from the remote disk. When a connection is made to another computer, an Exchange notice icon appears on the remote user's desktop. The name of the user making the connection appears in the Control menu. If for some reason the remote user does not want you to access his or her files and directories, that person can discon-

---

* An *IP address* consists of a series of four numbers separated by a period, such as 130.147.19.2. An *IP name* is a series of words separated by periods, such as nathan.iquest.com.

nect you and all other users or only certain users. Any files that are in the process of being transferred when the disconnect command is given will not be copied to the other computer.

Security is maintained by user-definable password for each level of functionality and enhanced with the ability to register users. Users may define what function each registered user can access with a unique password. For example, one registered user, through a specific password, may only be able to send files. A different registered user might only have remote-control privileges. This ensures that only preapproved, registered Timbuktu Pro users are allowed access on a specific system.

Stac Electronic's ReachOut

Stac Electronic's ReachOut remote-access software also works over Internet connections. If you're transferring only a small block of data from PC to PC, you can use ReachOut's Remote Clipboard feature. The Rapid-Sync feature lets you synchronize directories between connected PCs. ReachOut can save time by transferring only those parts of a file that have changed since the last time you connected. File transfers are very fast due to the product's use of Stac's compression technology.

With a script file you write, ReachOut lets you create connection icons and drop them onto your desktop for convenient access to remote machines. The package's PersonalFTP Server publishes your desktop to any FTP-enabled Web browser, making files available for remote access. You can specify who can access the host data, set access to drives and directories, limit log-on attempts, and track log-ons. ReachOut's SuperFTP Client, an industry-standard FTP utility, lets you drag and drop files and entire directories through ReachOut Explorer.

Traveling Software's LapLink

Traveling Software's LapLink for Windows 95 provides integrated remote access, remote control, file transfer, e-mail, and chat in a single package that works over the Internet, as well as over dial-up lines via modem, serial and parallel cables, and infrared links. Of particular interest to Internet users is that all data is encrypted for safe transfer between computers. Via an Xchange Agent, a single mouse click automatically connects and synchronizes information between your computers, allowing you to work with the latest versions of files on the local and remote computers.

A 30-day trial version of LapLink for Windows 95 Version 7.5 can be downloaded from the company's Web site at:

http://www.travsoft.com/download/proddownload.htm

Other Remote-Access Products

Among the other remote-access products that work over the Internet is Microcom's Carbon Copy/NET, which takes full advantage of Microsoft Explorer 3.0's support for ActiveX technologies to provide access to application software and files residing in PCs connected to the Internet. Carbon Copy/NET can even be configured to launch with Microsoft Internet Explorer.

Other remote-access products that enable you to exchange directories, files, and messages over the Internet include Symantec's pcAnywhere, Avalan Technology's Remotely Possible/32, and McAfee Associates' Remote Desktop 32. Each of these products has a particular strength. When connected on the Internet with Symantec's pcAnywhere, your data is encrypted and downloaded files can be scanned for viruses. Avalan Technology's Remotely Possible/32 not only provides encryption, it supports several read-only remote sessions at the same time, making it particularly useful for training applications. McAfee's Remote Desktop supports audioconferencing, allowing users at each end to converse as they view a screen to hash out a problem.

Whiteboards

In its purest form, whiteboarding offers the means to collaborate with friends or colleagues over the Internet without the need for high-bandwidth connections and the hardware to handle audio- or videoconferencing. One such product is TalkShow, which was originally developed by Future Labs but is now offered by Quarterdeck through its acquisition of that company.

Quarterdeck's TalkShow

TalkShow has two components: a client that lets you join in a conference and a server that lets you join, create, or host a conference. TalkShow offers all the basic whiteboard features, including freehand drawing and high-lighting tools and a full text-editing capability. You can set the font size, color, and type, and even add notes to any document. The notes are put on a separate layer so they won't impact the existing text. You can even pre-load slides to be discussed in a meeting. The application also includes OLE 2.0 integration, making document collaboration more useful because you can link to and update live documents.

With TalkShow, you also can share any Windows-based application with other users on the network. For instance, you can share and collabo-rate on a live file in an application such as Microsoft Word or Excel with another user on the network, even if that user does not have Word or Excel installed. The software lets remote meeting members perform live editing by pointing to or highlighting items on the documents that need clarification or changes. With support for OLE 2.0, users can drag and drop images directly onto the TalkShow whiteboard, view multiple objects, and make changes to documents.

To schedule and set up a virtual meeting, conference participants use a Web browser to access the Future Labs Conference Center and sim-ply enter a conference name, user name, and password. Future Labs also makes the necessary local IP (Internet Protocol) connections so users don't have to know the IP address of other users they wish to conference with.

Attachmate's OpenMind

Another product that offers collaborative workspace is Attachmate's OpenMind. In addition to providing everyone with the same view of information so they can develop documents jointly, OpenMind allows you to collaborate with others electronically with features such as forum-style conferencing. This lets everyone interact with messages in real time, like a chat facility. Unlike simple chat tools, however, everyone participates in threaded discussions in various forums similar to the way public news-groups operate (see Chap. 10). But unlike newsgroups, OpenMind forums are private.

OpenMind comes with Attachmate's Emissary browser. You can attach a Uniform Resource Locator (URL) in the collaborative workspace where other users can see it as a hypertext link. You can also create your own local HTML pages or map Web pages to your workspace for shared viewing.

OpenMind stores information in native formats, allowing you to copy all or parts of that information from a remote computer and paste it into other applications. It includes over 150 viewers that support most file formats including Word, WordPerfect, Excel, Lotus 1-2-3, and PKZIP archive files. Without starting any additional applications, you can print, copy, and paste most application files, with formatting and graphics intact. OpenMind can also launch other applications and exchange information with e-mail systems.

VocalTec's Internet Conference

VocalTec's Internet Conference offers a whiteboard from which multipoint connections are made and other activities are launched, such as chat, file transfer, joint Web page visits, private conference rooms for online meetings, and application sharing via OLE.

The program's conference-enhancing capabilities include color coding of each participant's annotations, the listing of participants' names at the bottom of the screen, and the ability to save an entire conference for future reference. Images are kept separate from annotations, so the annotations can be cleared without wiping away an image. Annotation tools include a pen, highlighter, text tool, and eraser. Drawing tools include those for creating lines, rectangles, ellipses, and diamonds. Internet Conference also provides screen-capture and cut-and-paste facilities and a remote screen synchronization tool.

Application sharing is accomplished through OLE, specifically, by cutting and pasting a file into Internet Conference from an external OLE-compliant application or by using the program's Insert-Object menu command. Once the file is inserted into an active window, all conference participants can view it. Any participant who has the native application loaded on his or her own computer can double-click on the file. This causes the application to load, after which the file can be edited. Only one participant at a time can edit the file. Because the program does not update remote participants' screens until the editor exits the OLE application, the other participants are not able to watch the editing process.

Internet Conference also offers a basic chat capability. Participants can type or paste comments into the data entry window, which appears at the bottom of the whiteboard. A larger, scrollable window appears on top displaying the entire chat with the authors' names to the left of their comments.

The program lets you transfer files among all participants or just to selected participants. Optional compression can be invoked to enhance performance.

Future versions of Internet Conference will include integrated audio via the company's Internet Phone (see Chap. 2). This will enable participants to talk to each other while they collaborate within the whiteboard. Future versions of Internet Conference and Internet Phone will include videoconferencing capabilities.

A 14-day evaluation copy of Internet Conference can be downloaded from VocalTec's Web site at:

http://www.vocaltec.com/conference/iconf_dnld.htm

Creative Software's CollabOrator

Creative Software Technologies offers a conferencing package that includes such collaboration basics as application sharing, whiteboard, chat, file transfer, and slide presentations. In the presentation mode, you can link several BMP files into a single presentation. Both parties can flip through and annotate the individual slides.

However, the company's entry level product, CollabOrator System 1000, allows only two-party communication and does not have an audio capability. The more advanced capabilities are offered in CollabOrator System 9000, which includes audio and video multipoint conferencing over the Internet.

A midlevel offering is CollabOrator System 2000, which is a fully integrated Internet telephony and data collaboration package that allows two users to converse with each other, as in a regular telephone conversation, while simultaneously making a presentation or developing ideas on a whiteboard and transferring files. However, this product relies on a dedicated codec card to provide full-duplex voice. The product's CallThru facility allows users to make long-distance voice calls over the Internet to regular telephones anywhere in the world.

Unfortunately, the company does not offer a trial version of its entry-level product. Given the sophistication of competitive offerings and the willingness of so many other vendors to offer beta and demonstration versions of their products, interested users might do better looking elsewhere for collaborative software rather than risk disappointment after the purchase.

Other Whiteboard Products

There are several other whiteboard products available. FTP Software's GroupWorks, for example, provides peer-to-peer connectivity over the

Internet, enabling project collaboration in the form of task assignments, document distribution, links to external data, and discussions among project participants.

Discussion and Presentation Tools

The whiteboard is usually thought of as a shared space within which users can view documents, drawings, and images while they take turns annotating them as they chat back and forth. However, some vendors offer discussion forums as the shared space within which project collaboration takes place. These tools allow information to be organized into easily accessible forums. You can use forums for a variety of discussion applications including product development, project management, corporate help desks, management roundtables, and customer service. You can also use this kind of software to disseminate information such as company announcements, policies and procedures, human resources information, forms, and reports. Entrepreneurs can also make use of this software to collaborate with partners at remote locations, using the Internet as a virtual office.

Netscape's Collabra Share

An example of this type of product is Collabra Share, originally developed by Collabra Software, but which is now offered by Netscape Communications. Collabra Share provides the capability to set up private forums organized around specific projects (Fig. 9.1). These forums can be moderated or unmoderated. In a *moderated forum,* an administrator or supervisor reviews any new messages before they become public. You can choose to moderate all messages or just the ones submitted anonymously.

The discussions are *threaded,* meaning that messages are arranged in a logical order so that members can follow the context of the discussion. As members of a forum, e-mail users are not only able to receive messages, but they can also create replies that go to the correct thread position.

Collabra Share also gives each user personal organization tools for customizing discussion information. For example, you can create groups of

Figure 9.1
With Collabra Share,
you can set up pri-
vate forums around
specific projects,
issues, or organiza-
tional groups.

forums that let you organize them in a folderlike system. You can also create a personal hotlist of messages. A search engine that uses boolean parameters lets you look for any word or string that may appear in the subject field and body of the message. The returned list of messages can be viewed individually within the search result list or opened within the context of the actual forum to which the message belongs.

Collabra Share also offers connectivity to a range of other products and technologies with Collabra Share Agents. These agents provide links to other information sources such as Usenet Newsgroups, Lotus Notes, online newsfeeds, e-mail participants, and replication across the entire company or to other partners. At this writing, Netscape is in the process of integrating Collabra Share's group discussion features into the next full release of Navigator. The result will be a Web browser that lets users engage in group discussions whether they are on a corporate intranet or the greater Internet. Collabra Share will continue to be available as a stand-alone product.

A 90-day evaluation copy of Collabra Share is available for download from Collabra Software's Web site at:

http://www.collabra.com/products/trial.htm

PictureTalk's PictureTalk Communicator

A useful tool for delivering presentations is PictureTalk's Communicator software, which offers real-time visual communication to multiple users over business networks or the Internet. While you're in Communicator, whatever you can put on your screen is what everyone else sees. The PictureTalk Communicator allows any number of users to join a meeting. It also allows each user to join more than one meeting.

The company's client software continuously captures and compresses screens from the transmitting PC, sending them over any network to the PictureTalk Conference Server, which routes the images to all designated recipients. The screen image is continuously updated, with compression rates varying according to the CPU and graphics power of the participating systems.

The company's client and server programs work with PC, Macintosh, and UNIX workstations and any installed application software—for example, letting a PowerBook user in New York conduct a presentation for a PC colleague in Atlanta and a UNIX user in Chicago.

The PictureTalk Conference Server manages and provides security for PictureTalk meetings via a Web server. The company also offers PictureTalk Place, a virtual meeting place similar to an online forum for Internet visitors to conduct conversations and test-drive the PictureTalk software.

The PictureTalk Communicator is free. You can download a copy in either Windows, Mac, or UNIX versions from the company's Web site at:

http://www.picturetalk.com/PictureTalk/htdocs.ptk/dload.html

It works with most popular Web browsers, including those from Netscape, Microsoft, Oracle, and Spry.

Application Sharing

Application sharing lets conference participants view shared documents and use the application's native tools to edit, annotate, and mark them up. The way in which application sharing is implemented differs according to vendor.

Microsoft's NetMeeting

NetMeeting, which comes with Microsoft's Internet Explorer 3.0 for Windows 95, provides a number of facilities that enable and support application sharing. The software enables participants to do such things as edit a Word document, review a PowerPoint presentation, annotate a FreeHand drawing, or play what-if with an Excel spreadsheet. Together, participants can look at the same screens as you delve into data in a corporate or CD-ROM database or browse for information on the World Wide Web. You can copy documents and images from any source into the whiteboard and mark them up any way you want.

To implement the application-sharing feature, you simply click on the Share icon and pick an application. This sends the application's screen image to all participants. At that point, remote users have access to the program's tools and any documents you open, even if they do not have a copy of the application running on their computers.

NetMeeting defaults to Work-alone mode, which lets participants view but not control the program. Once the host participants selects Collaborate mode, any participant can take control of the application just by clicking on the program's image. This lets the remote participant use any function of the program, including Copy, Paste, Save, Delete, and Exit. NetMeeting identifies who is controlling the application by placing the user's initials next to the moving cursor.

NetMeeting's whiteboard, which looks like Microsoft Paint, can be launched through an icon or menu command. This triggers the application on all participants' computers. Since the whiteboard is *object-oriented* (versus *pixel-oriented*), you can move and manipulate its contents by clicking and dragging with the mouse. In addition, you can use a remote pointer or highlighter to point out specific contents or sections of shared pages. The whiteboard supports multiple pages and has a synchronization option that keeps all participants on the same page. When one participant exits the whiteboard, all others retain the contents until they exit.

NetMeeting offers an audio capability that allows two participants to have a voice conversation during a conference. (Up to five users can participate in a data conference; more if the conference is established through a network-based conferencing service. NetMeeting even includes a tool for connecting to and scheduling the use of network-based conferencing services such as those offered by AT&T, MCI, ConferTech, and others.) The other participants will not hear the conversation. Fortunately, NetMeeting also includes a text-based chat capability that allows all participants to join in a conversation. As participants join the conference, NetMeeting lists them in a program window. The window identifies which two have voice capabilities and describes their application-sharing status.

NetMeeting activates its chat utility on all systems when any conference participant initiates a chat session. The interface has two panes: a large, scrollable window on top that tracks the conversation, and a smaller window on the bottom for entering text. Since the program does not send messages until you press the Enter key, you can edit your remarks before sending them. Any participant can save the conversation. On exiting, the program prompts you to save.

NetMeeting's file transfer capability enables you to send a file to a specific person or all participants in a conference. You can right-click on a person to send them a file, or you can drag a file into the NetMeeting window and have it sent automatically to each participant in a conference. The file transfer occurs in the background, allowing everyone to continue using other NetMeeting facilities.

NetMeeting's shared clipboard allows you to exchange the contents of the clipboard with other participants in a conference. This allows, for example, a user to copy information from a local document and paste it into the contents of a shared application as part of a group collaboration. This capability provides seamless exchange of information between shared applications and local applications, using the familiar Cut, Copy, and Paste operations.

NetMeeting supports the T.120 standard recommended by the International Telecommunications Union (ITU).* This means NetMeeting is interoperable between applications and across platforms. At this writing, NetMeeting works with over a hundred products from different vendors.

A copy of NetMeeting can be downloaded from Microsoft's Web site at:

http://www.microsoft.com/netmeeting/download/

ForeFront Group's RoundTable

The ForeFront Group offers a conferencing product called RoundTable that relies on the Web to bring participants together. Typically, a moderator will notify participants of a meeting time via e-mail or a telephone

---

* T.120 is actually a suite of dial-up and network-based data conferencing protocols that specifies how applications may interoperate with (or through) a variety of network bridging products and services that also support T.120.

call. At the scheduled time, users will browse the conference Web page which lists live sessions, each represented by a hypertext link. As new meetings are created, the Web page is updated accordingly. Clicking on a session gets the attention of the server, which sends a message back to the browser telling it to load RoundTable as a helper application. The server then establishes the connection and drops the user into the meeting. RoundTable becomes the user interface, and the browser recedes into the background.

The RoundTable meeting window is organized around three panes—the Chat Panel, the Meeting Roster, and the Canvas workspace. The *Chat Panel* is used for typing, sending, and viewing text-based chat messages between conference participants. The *Meeting Roster* displays the list of participants at the meeting. The *Canvas* provides a workspace for sharing multimedia information. Participants simply drag and drop or Cut/Copy and Paste images, documents, URLs, audio, video, or any other type of file into the canvas to share it with everyone else.

RoundTable does not, however, let multiple parties manipulate the file, as in true application sharing. If participants want to change something being reviewed, they have to download it and work on it using a local application. True application-sharing capabilities are planned for a future release, as is real-time audio support, which would do away with the need for a separate teleconference.

The RoundTable client software is available free from ForeFront's Web site at:

http://www.ffg.com/

After downloading and installing the software, you can visit Fore-Front's RoundTable Conference Site to join a meeting (Fig. 9.2):

http://www.ffg.com/cgi-bin/http2conf

Visual Components' Formula One/Net

Some products allow you to share an application through a Web browser. Visual Components, for example, offers an Internet spreadsheet component called Formula One/NET. With Formula One/NET, you can embed live Excel spreadsheets and charts directly in a Netscape Navigator 2.0 window.

Formula One/NET is not a viewer or player; rather, it is a fully functional spreadsheet application implemented as a plug-in for Netscape Navigator. With Formula One/NET installed, a user can view, edit, and interact with worksheets that include live charts, links to URLs, formatted

Figure 9.2
To join a conference, you can access Fore-Front's RoundTable Conference Site on the Web. There, you can reload the Web page to display an updated list of meetings and participants.

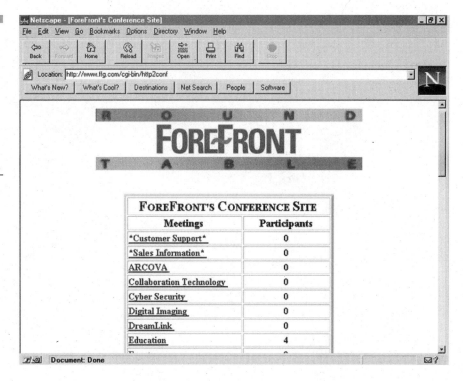

Figure 9.2
To join a conference, you can access Fore-Front's RoundTable Conference Site on the Web. There, you can reload the Web page to display an updated list of meetings and participants.

text and numbers, formulas, and buttons, and controls. Once a spreadsheet has been sent to a user's Navigator window, it is a local copy. The spreadsheet can be saved locally on the user's hard drive or on a networked volume.

Formula One/NET is not shareware, but an extension of the Formula One product family. At this writing, Formula One/NET is only available for Windows platforms. You can download a free copy of Formula One/NET from the company's Web site at:

http://www.visualcomp.com/f1net/download.htm#now

Calendaring and Scheduling

For many people who use personal information managers (PIMs), it would be helpful if they could share their calendars and schedules with others. This would improve efficiency and productivity. A secretary, for example, would be able to access your PIM on the Web to avoid scheduling potential appointment conflicts when you are away from the office. At the same time, you could update your calendar while traveling and, via the Web, share the new information with anyone who needs it.

Now Software's Up-to-Date and Web Publisher

Now Software provides a way to improve information sharing between users of its personal information manager (PIM). The company offers two products that, when used together, allow users to drag schedules and other important data from intranet Web pages and drop them into electronic address books and calendars created in Now's PIM.

The first product, Now Up-to-Date Web Publisher, lets organizations convert calendars and contact databases to HTML format, generating Web calendars and address books without the need for programming. The second product, Now Up-to-Date, is an upgrade to Now's PIM and is designed to grab addresses and schedules embedded in those Web pages. The products are available in both Apple Macintosh and Microsoft Windows versions.

You can download a free 30-day trial copy of both Now Up-to-Date Web Publisher and Now Up-to-Date from the company's Web site at:

http://www.nowsoft.com

The company also offers calendar and address book plug-ins for Netscape. With the AboutTime plug-in, you can view browsable calendars in a Netscape Navigator window. With the AboutPeople plug-in, you can view searchable address books in a Netscape Navigator window. Working examples and beta copies of these plug-ins are also available at the company's Web site.

Other Calendaring and Scheduling Products

Other PIM-related products offer varying degrees of operability over the Internet. NetManage's ECCO Pro 3.0, for example, only launches a connection to the Internet and allows you to add URLs to a contact database. Others, such as GoldMine Software's GoldMine for Windows 95, provide the ability to manage Internet e-mail, link messages to contacts, and synchronize contact databases via the Internet. These and other Internet features will become standard in the next generation of information managers.

Online Resources

The following table provides the Web links of the major providers of collaborative products that work over the Internet. In most cases, you can download copies of their software before deciding whether to buy the commercial version. The Web pages also contain such information as platforms supported, system requirements, product features, installation instructions, and troubleshooting advice. Since the technology is moving rapidly, you may want to check these Web pages periodically for the latest developments.

Developer	Product	Web page or FTP site
Attachmate Corp.	CrossTalk, OpenMind	http://www.attachmate.com
Avalan Technology	Remotely Possible/32	http://www.avalan.com
Creative Software Technologies Pty	CollabOrator System 1000, 2000, 9000	http://www.cst.com.au/
Digital Equipment Corp.	AltaVista Forum	http://www.altavista.software.digital.com/ products/forum/nfintro/htm
Farallon Computing	Look@Me, FlashNotes, Timbuktu Pro	http://www.farallon.com
ForeFront Group	RoundTable	http://www.ffg.com/
FTP Software	OnNet for Windows, GroupWorks	http://www.ftp.com
Gigatron Software Corp.	Note It!	http://www.gsclion.com/
GoldMine Software	GoldMine for Windows 95	http://www.goldminesw.com
InSoft	CoolTalk, CoolView	http://www.insoft.com
Lotus Development Corp.	Organizer 97 Web Calendar	http://www.lotus.com
McAfee Associates	Remote Desktop 32	http://www.mcafee.com
Microcom	Carbon Copy/NET	http://www.microcom.com/
Microsoft Corp.	NetMeeting	http://www.microsoft.com/netmeeting
NetManage	ChameleonNFS/X,	http://www.netmanage.com
Netscape Communications Corp.	Collabra Share	http://www.netscape.com
Now Software	Now Up-to-Date Web Publisher	http://www.nowsoft.com
O'Reilly & Associates	WebBoard	http://www.ora.com

PictureTalk	PictureTalk Communicator	http://www.picturetalk.com/
Quarterdeck Corp.	TalkShow	http://www.quarterdeck.com
Stac Electronics	Reachout	http://www.stac.com
Symantec Corp.	pcAnywhere	http://www.symantic.com
Traveling Software	LapLink	http://www.travsoft.com
Tribal Voice	PowWow	http://www.tribal.com/powwow/
Visual Components	Formula One/NET	http://www.visualcomp.com
VocalTec	Internet Conference (Personal and Professional Editions)	http://www.vocaltec.com

NOTE: This information, as well as updates, can be found at http://www.ddx.com/mgh.shtml.

Conclusion

More and more new applications offer Web connectivity, allowing geographically dispersed users to simultaneously share information, edit documents, see each other's desktops, swap files, and even converse with each other as they collaborate on a project to meet a deadline.

Gigratron Software's Note It!, for example, is a software package for annotating virtually any kind of document or file. It accommodates hundreds of different file formats, including word processing documents, Internet Web pages, graphics, faxes, and spreadsheets. It allows you to mark up, revise, and highlight documents or other files without altering the original (Fig. 9.3). You can even attach sticky notes to a document containing comments and suggestions. You can date-time stamp annotations, attach voice notes, and integrate OLE-supported objects such as video and spreadsheets into the Note It! document. Designed for Windows 95, Note It! is also MAPI-compliant. It includes a built-in e-mail link, which provides the ability to receive files over LANs as well as the Internet.

Even conventional applications are starting to become Internet-enabled. For example, Lotus Development Corporation's SmartSuite 96 takes advantage of the Internet in several ways. From within Word Pro, for example, users can open and save files to a File Transport Protocol (FTP) server without having to access a helper application registered with their Web browser. In addition, Freelance Graphics users can publish their presentations on the Internet in HyperText Markup Language (HTML) format and virtually any Web browser can be used to view them.

Figure 9.3
Gigratron Software's
Note It! allows you to
share documents and
images and mark
them up using a vari-
ety of tools.

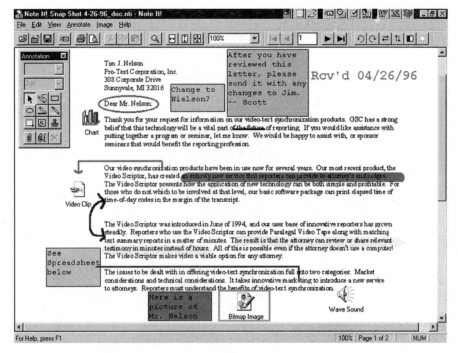

In the development of its Office 97 application suite*, Microsoft has given top priority to the integration of applications with the World Wide Web. Office 97 applications such as Excel and Word integrate seamlessly within a Web browser that supports a feature called Active Documents. From within an Active Documents—compliant Web browser, you can share and edit Excel or Word files.

In addition, Office 97 users will be able to create and save documents from any Office application in native HTML format. You can publish Access databases in HTML format, complete with the ability to query and update the data dynamically. Within Office apps, you will be able to create hyperlinks to and from any Office, HTML, or third-party file on a Web site or file server. An Office Web toolbar with forward and back buttons will navigate between linked documents.

A noteworthy addition to the Office suite is Outlook, which is a personal information manager that provides a single, integrated environment for managing tasks, schedules, files, and e-mail. Via Microsoft's Exchange

---

* The Office 97 application suite includes Access, Excel, FrontPage, PowerPoint, Publisher, and Word, plus Outlook, an Internet-enabled personal information manager (PIM).

Server, users can share Outlook calendars, contacts, or tasks that are accessed in a Microsoft Exchange Public Folder. With its support of POP3 and SMTP, users can create hyperlinks to internal or external Web sites, calendar items, contacts, and tasks within an Outlook e-mail message.

In the not-too-distant future, all major word processor, spreadsheet, graphics, database, personal information manager, and scheduler programs will support true multiuser access and collaboration over the Internet. Indeed, the ability of such products to work over the Internet may be among the key selection criteria for new purchases, especially among corporations where such trends as decentralizing operations and employee telecommuting are on the increase. With increasing reliance on the Internet, such companies must have products that facilitate employee communication and productivity, regardless of the distance between collaborating participants.

Collaborative software is not just for big companies; it offers benefits to individuals, too. A lone entrepreneur working out of his or her home, for example, can use Internet-enabled software to facilitate communication or project completion with a corporate client. Having the same tools as big companies can even enhance your ability to obtain their business. With the right software, you would already be equipped to integrate into the company's operations and workflow patterns. Potential competitors who are not equipped with the means to collaborate and who must still rely on face-to-face meetings to get work done are increasingly looked upon as a burden on staff time and company resources. Such people will be avoided whenever possible.

If this is your situation, you will want to pay attention to collaborative tools that adhere to industry standards such as T.120. This will ensure that you can communicate and collaborate with various corporate clients, regardless of what specific vendor's application they happen to be using. Since corporations put a premium on interoperability when they choose software products for multiuser networking, you will be able to fit into their work environment better if you put the same emphasis on standards when selecting this kind of software.

CHAPTER **10**

Participating in Newsgroups

Introduction

Newsgroups can be described as electronic discussion groups or bulletin boards that operate over the Internet as a service known as *Usenet.* Depending on who is doing the counting, the number of newsgroups is estimated to be between 15,000 and 20,000. Ostensibly, each is devoted to a particular topic, but there is considerable overlap in topics across different newsgroups. These newsgroups are intended to provide a forum for informal communication among people with shared interests.

Newsgroups enable you to participate with others in a remote dialogue by posting and reading messages on topics of mutual interest. Discussion groups support multiple conversations, or *threads,* on a given subject, displaying postings in the context of the prior discussion. This allows you to follow an entire discussion from start to finish, even though you may have joined the discussion well after it started. Within the same newsgroup there may be dozens of threads in progress at the same time.

Not all newsgroups are publicly accessible. Many companies, associations, professional groups, research institutions, and government agencies operate their own news servers to disseminate organizational and membership news and to provide a vehicle for group collaboration. For example, a corporation can install a private news server on its network to facilitate collaboration and communication within workgroups, across departments, and between its personnel in remote offices around the world. Through the news server, secure discussion groups can be used by project teams for such purposes as developing marketing plans for new products, discussing technical solutions that can improve customer service, or just brainstorming better ways of working together.

Many private news servers support the use of standard MIME types and image formats, including GIF and JPEG. Attaching documents in a variety of multimedia formats greatly expands the capability of a discussion group to serve as a collaborative communications tool. Colleagues can download documents sent to the group, mark them up, and send them back. In supporting MIME attachments, any number of common document formats can be attached, making private corporate newsgroups a convenient medium for workgroup collaboration.

There are also many public newsgroups that have been set up for sharing so-called binary files. These are not ordinary text files; rather, they are files that include attached images that have been encoded and split into several parts. This allows the posting of images into the newsgroups along with text files. Some newsgroups—particularly those with *binary* or *pic-*

ture in their names—have as their primary purpose the posting of images for downloading by others. A common form of image posting uses the uuencode (UUE) format. To handle these binary files, you will need a newsreader that supports this decoding and encoding method.

Whether public or private, newsgroups are accessed with a *newsreader,* special software that offers various features to help you find, organize, and participate in newsgroups. Using the newsreader you can call up a complete listing of all newsgroups, browse through the ones that may be of interest, and subscribe to the ones in which you have a continuing interest. Of course, you can also unsubscribe from a newsgroup at any time. The newsreader can also help you follow the various threads, reply to articles posted by others, and post articles of your own.

If you are accessing a particular newsgroup for the first time, you should read any introductory messages posted before attempting to participate. For example, sometimes there is a FAQ (frequently asked questions) file associated with the newsgroup that you can search through to get updated on the topics that have been discussed at length so you don't waste everyone's time on issues that have already been covered.

After creating an article on your local computer, you upload it to a Usenet computer, which is most likely your own Internet service provider, which relays it to other computers until all the participating machines worldwide have received it.

Unless screened by a moderator, articles are automatically posted to the designated newsgroup, where they may be read by anyone. Depending on how busy the newsgroup is, it may take an hour or longer for your article or message to appear. If a newsgroup is moderated, it may take longer to post the articles, perhaps a day or two, depending on the moderator's workload. The moderator may insist that the discussion remain focused on a particular topic. Articles that are not germane are simply not posted.

About Usenet

Usenet—short for *users' network*—was conceived in 1979, when two Duke University graduate students in North Carolina, Tom Truscott and Jim Ellis, got the idea of linking computers together to exchange information within the UNIX community. Steve Bellovin, a graduate student at the University of North Carolina, put together the first version of the news software. As the Internet grew, other individuals at other schools improved and expanded the news software until, today, it runs on most types of computers from PC to mainframe. Usenet now accounts for about 10 percent of the total communications traffic on the Internet.

Like the World Wide Web (WWW) and Internet Relay Chat (IRC) discussed in previous chapters, Usenet is a service that operates over the Internet. Usenet consists of all the computers whose administrators have agreed to exchange specially formatted messages called *articles*.

As with the World Wide Web, Usenet has no central authority. The transmission of Usenet news relies on the cooperative and voluntary efforts of system administrators, who may or may not be subsidized in whole or in part by business, government, or academia. Many Internet service providers support this effort as well.

These administrators maintain the news software, monitor disk space, assist in the implementation of new newsgroups, and help end users configure their newsreaders. On small systems, this person may simply be the overall system administrator. At larger computer sites, the system administrator may designate someone else as the news administrator. Handling news-related issues may be one of several overall job responsibilities.

If you have questions about configuring your newsreader software, you should contact your news administrator. And if you want to read a newsgroup which is not currently available on your system, your news administrator is the person who can either arrange to receive that newsgroup or explain why it is not available.

Each Usenet computer maintains a database of articles, which it keeps up-to-date by exchanging articles periodically with its neighbor systems on Usenet. Periodically, your news system receives batches of articles from its Usenet neighbors; this may happen as frequently as every few minutes or once a day. Similarly, your news system periodically sends locally created articles to its neighbors. It may also pass its incoming batches on to some of its neighbors. Finally, your news system periodically (usually once each night) removes old articles to make room for new ones. How long articles are kept can vary from one newsgroup to another and depends on the amount of disk space available and the perceived value of the newsgroup.

To convey news, Usenet relies on two major transport methods, UUCP (UNIX-to-UNIX Copy Protocol) and NNTP (Net News Transport Protocol). The former is a network of UNIX hosts and workstations linked by high-speed lines and modem-based access connections. This network is not really a part of the Internet, except where it provides interconnections via e-mail gateways. Therefore, users incur the normal charges for telephone calls. The latter, NNTP, is the primary method for distributing news over the Internet.

With UUCP, news is stored at a feed site in batches until a neighbor site calls to receive the articles, or the feed site calls on a scheduled basis. A list of newsgroups that the neighbor wishes to receive is maintained at the feed site. Some systems compress these batches prior to sending them out, which reduces the transmission time necessary for a relatively heavy newsfeed, in turn reducing long-distance charges.

NNTP, on the other hand, offers more flexibility in how news is sent. Although the traditional store-and-forward method characteristic of UUCP is still used, many news servers have dedicated connections with their neighbors, allowing news to be sent nearly instantaneously. Depending on the available bandwidth, they can handle dozens of simultaneous feeds, both incoming and outgoing. By attaching unique IDs to articles, systems are prevented from receiving multiple copies of the same article from each of their neighbors.

NNTP also supports the creation and maintenance of private discussion groups at corporate sites. Most newsreaders are based on NNTP and a growing number are adding support for the Secure Sockets Layer (SSL) protocol to safeguard communication between clients and news servers. SSL provides encryption, server authentication, and message integrity, which address corporate security concerns.

The Newsgroup Hierarchy

Usenet consists of all the computers on the Internet that exchange articles tagged with one or more universally recognized labels, called newsgroups. Newsgroups are organized in a tree structure, falling under broad categories and subcategories, as appropriate. There are seven major newsgroup categories:

comp. Includes over 780 individual newsgroups covering a wide range of computer-related topics such as software and hardware, operating systems, robotics, viruses, research, human factors, and jobs.

misc. Includes about 120 individual newsgroups that are not easily classified into any of the other headings or which incorporate themes from multiple categories. The subjects are quite varied and include health and fitness, education, consumer issues, investments, taxes, job hunting, law, and off-beat topics such as survivalism.

sci. Includes about 168 individual newsgroups on science-related topics. Subjects include aeronautics, agriculture, anthropology, archaeology, energy, physics, psychology, and space.

soc.   Includes about 215 individual newsgroups addressing various social issues, including atheism, feminism, history, culture, politics, religion, and human rights.

talk.   Includes about 28 individual newsgroups that are mostly opinion-oriented. Topics include abortion, the environment, euthanasia, politics, and religion.

news.   Includes about 25 individual newsgroups that are concerned with the news network, including its administration and maintenance. This is where individuals propose the creation or modification of newsgroups. There is also a newsgroup devoted to newsreader software.

rec.   Includes over 600 individual newsgroups related to hobbies and recreational activities. Topics include animals, antiques, art, aviation, boating, beer drinking, hunting, pets, skiing, travel, and woodworking.

There are also several alternative and special-purpose newsgroup categories, such as

alt.   Includes about 2700 individual newsgroups covering a broad range of topics, including aliens, astrology, bigfoot, freemasonry, satanism, sex, show business, skinheads, spanking, suicide, supermodels, torture, and yoga.

biz.   Includes about 50 business-related newsgroups on such topics as books, computing, and jobs. Many of these newsgroups are company- or product-specific.

us.   Includes about 25 newsgroups related to the United States covering such topics as taxes, jobs, and legal issues. For some reason, 13 of these newsgroups are related to South Carolina, including Charleston, Columbia, Florence, Lancaster, and Rockhill.

Virtually anyone can propose the creation of a newsgroup. However, this shouldn't be done without first doing extensive research. A group might already exist that you are just not familiar with. If you are satisfied that one doesn't exist, the next step should be to start a discussion in a related group about what kind of newsgroup you want to create. This will give you an idea of how much support there is for your idea. If you are encouraged, the next step is to formalize your proposal so it can be voted upon. The specific procedure to follow is described in App. B.

Creating a new newsgroup is much easier in the alt hierarchy than one of the seven standard hierarchies, so a new newsgroup often appears there first with the intention of generating enough interest to justify creating a newsgroup in one of the standard hierarchies. You can get more information on creating a new alt group by opening your newsreader and going to one of these newsgroups:

alt.config

news.groups

news.answers

Getting Started

You can get started participating in newsgroups by browsing through the various lists and sampling their contents. First-time users should also check out the following newsgroups periodically:

news.announce.important.   Contains important messages of interest to all Usenet users, including announcements of newly created newsgroups

news.announce.newusers.   Contains a standard set of articles with general information about Usenet and guidelines for network etiquette

news.answers.   Contains copies of articles which are periodically posted to various newsgroups, providing answers to various frequently asked questions (FAQs).

news.newusers.questions.   Contains questions from new users and answers from more experienced users.

With so many newsgroups to choose from, it is sometimes difficult to find a newsgroup that coincides with your interests. If you are using a Web browser—Netscape Navigator, for example—the easiest way to find newsgroups is to go to Window, open the Netscape News window, then select Show All Newsgroups from the Options menu. This will return a long list of newsgroups in alphabetical order, with main headings indicated with a file folder icon (Fig. 10.1) and a number in parentheses indicating how many individual newsgroups are associated with that heading.

Another way to find topics of interest is to use a query form that allows you to search newsgroups by key word. For example, there is an advertiser-supported query form provided by Deja News Research Service (Fig. 10.2) which can be found on the Web at:

http://www.dejanews.com/forms/dnq.html

Deja News allows you to enter search parameters and presentation formats, including key words matched (all or any); Usenet database (current

Figure 10.1

A listing of all news-
groups, as shown by
Netscape Navigator's
newsreader. Not only
are the number of
newsgroups indi-
cated next to the
group folder, but the
number of unread
and total messages
associated with each
newsgroup are
shown. You can sub-
scribe to a news-
group by clicking its
associated checkbox.
This places a check
mark in the box. To
unsubscribe, click on
the checkbox again.
This removes the
check mark.

or old); number of hits per page (30, 60, 120); hit list detail (concise or
detailed); hit list format (listed or threaded); sort by score, newsgroup, date,
author; article date bias (prefer new or prefer old), and article date weight
(some, great, none). You can even create a query filter to limit your search
by newsgroup, date, or author.

The results of your search will be returned, sorted by the parameters
you set in the query form (Fig. 10.3). A search on *Parkinson's Disease*, for

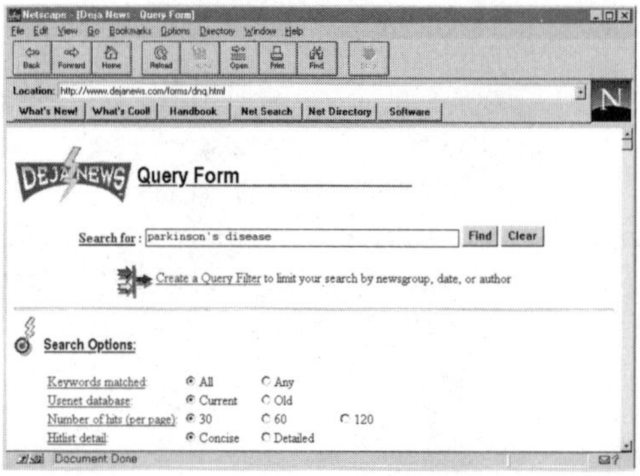

Figure 10.2
The Deja News query
form allows you to
search newsgroups
by a number of differ-
ent parameters and
presentation formats.

Figure 10.3
Deja News query results for the key word *Parkinson's Disease*. Clicking on the links brings you to the articles.

example, resulted in a list of 685 articles being returned. The list spanned all the newsgroups in which the term appeared in various posted articles. By clicking on the links, the articles are retrieved for viewing in your browser.

Another Usenet search tool is the Liszt of Newsgroups which can be found on the Web at:

http://www.liszt.com/cgi-bin/news.cgi

Still another way to search for newsgroup content is to use an Internet search tool such as Alta Vista which is offered by Digital Equipment Corporation as a public service. Alta Vista is not only fast and comprehensive in its searches, it gives you the choice of initiating a search for material on either the Web or Usenet (Fig. 10.4).

If you are interested enough in particular newsgroups to want to visit them frequently, possibly with the idea of becoming an active participant, you can subscribe to them. This is something like bookmarking your favorite Web sites. The next time you open your newsreader, you can

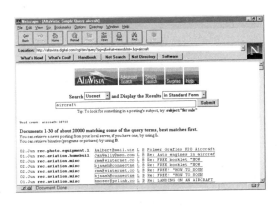

Figure 10.4
Alta Vista allows you to initiate key word searches on either the Web or Usenet. Search results can be displayed in standard, compact, or detailed form. Clicking on the links brings you to the articles.

call up only the list of the newsgroups you subscribe to so you can access them quickly. Later, you can unsubscribe to various newsgroups if you become dissatisfied or bored with them. Subscribing and unsubscribing are accomplished simply by checking and unchecking the boxes that appear next to the newsgroups when you call up the list of all newsgroups (refer back to Fig. 10.1).

After monitoring a newsgroup for awhile, you might want to participate by posting an article that is relevant to the topic of discussion. You create a news article just like you would an e-mail message. In your newsreader application, you typically select a menu item or icon called Post or To: News, which causes a pop-up window to appear (Fig. 10.5). After filling in the header fields, you type a message in the space provided and/or enter the location of an attachment. An attachment is a document you have created offline, that can be appended to the news article. Depending on the newsreader, you may even be able to create a customized signature that is added automatically to the end of your messages. This can be a humorous comment, a profound statement on the human condition, or something formal like your business or professional affiliation and contact information.

If all this sounds like e-mail to you, it is quite similar. Although similar to e-mail, articles are transmitted over the Internet via NNTP, rather than the Simple Mail Transfer Protocol (SMTP), which is used for most e-mail transmissions. Another difference is that articles are intended for public posting and comment, while e-mail is meant for personal communication.

Figure 10.5
Message composition window called from within Netscape Navigator's newsreader.

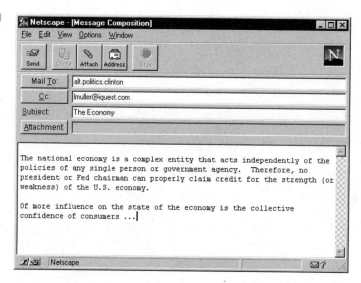

System Requirements

Newsreaders generally do not require very much in the way of computer resources. In most cases, a 80386-based PC will do. However, integrated products such as Attachmate Corporation's Emissary and NetManage's Chameleon, would perform better on a 80486 or Pentium. These products combine a newsreader with several other Internet applications, all of which are accessed from the same graphical user interface.

Memory requirements are modest: 4 MB RAM or higher for Windows 3.x, but 8 MB or higher is recommended, especially for Windows 95 users. Newsreaders do not take up much disk space. Upon installation, most stand-alone products take up less than a megabyte (MB). Again, integrated products are the exception: in compressed form, the files are about 4 MB, but they consume another 6 MB upon installation.

With regard to the monitor, a VGA display with 256 colors set for a resolution of 640 × 480 pixels is all you need for most newsreaders. Super VGA will give you a higher range of colors and a resolution of up to 1024 × 768 pixels, both of which are important in highly integrated products (i.e., Emissary). Setting the display for higher resolution gives you a larger work area from which to view more windows so you can work in multiple applications simultaneously. And, since integrated products offer connectivity to the highly graphical Web, having a display that combines higher resolution with more colors allows you to view Web pages with little or no dither.

To get your newsreader to work, it must be set up with the name of your news server (Fig. 10.6). A news server stores recent messages for all the newsgroups your Internet service provider tracks. You can find out the name of your news server simply by asking your Internet service provider or system administrator. Once you have this information, open Netscape and enter the news server's name in the Mail and News section of the Options/Preferences dialogue box. To read Internet news articles in CompuServe, after Go, type Usenet, and follow the prompts. In AOL, click on Internet Connection, then on the Newsgroups button.

To access newsgroups in the Microsoft Network (MSN), go to the MSN Directory and scroll down to Internet Newsgroups (Fig. 10.7). Or you can access Category from MSN Central and choose Categories. From there click on Internet Center, then Internet Newsgroups.

An interesting feature of Microsoft's newsreader is that it allows you to create shortcut icons for individual newsgroups, which are automatically placed on your Windows 95 desktop (Fig. 10.8). When you establish a con-

Figure 10.6
Setup screen of Trumpet Newsreader 1.0 showing the kind of information you will need before you can use your newsreader.

Trumpet Setup		☒
News Host Name	news	
Mail Host Name	mail.iquest.com	
E-Mail Address	nmuller @ ddx.com	
Full name	Nathan Muller	
Organization	The Oxford Group	
Signature file name	c:\winword\text\sigfile.txt	
POP Host name	iquest.com	
POP Username	nmuller Password ********	
	☐ Fetch Read-only	

Ok   Cancel

nection to MSN, you can then click on the shortcut icon of the newsgroup you want to access. This method of access lets you get to the newsgroups faster, without going through a lot of menu selections.

You must have an Internet connection established before running the newsreader. Run whatever Internet connection software you normally use for other Internet software (dialers, Trumpet, or other Winsock software) and establish your connection. Then simply double-click the newsreader program icon to open it. If you get an error message, it is probably because you do not have an Internet connection, and the newsreader will not run. Other sources of errors have to do with configuring the newsreader to recognize your news server or SMTP/POP3 e-mail server.

Figure 10.7
You can access newsgroups from the Microsoft Network (MSN) by going to the MSN Directory and scrolling down to and clicking on Internet Newsgroups. You can then access the full list of newsgroups or go to them alphabetically.

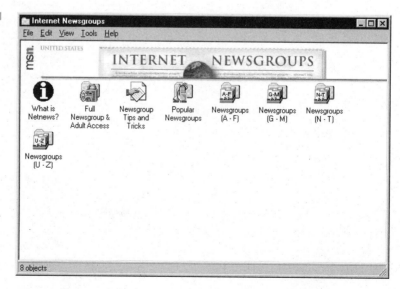

Figure 10.8
A shortcut icon has
been created for the
newsgroup
sci.archaeology (see
last icon in fourth col-
umn of icons) and
automatically added
to the desktop for
easy access the next
time an MSN connec-
tion is established.

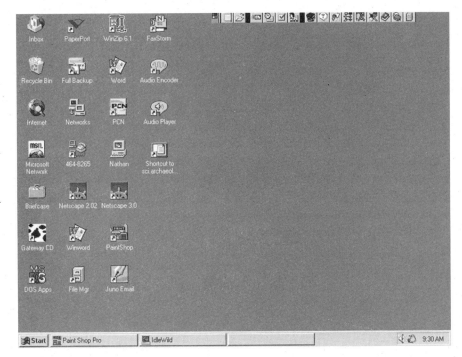

Figure 10.8
A shortcut icon has
been created for the
newsgroup
sci.archaeology (see
last icon in fourth col-
umn of icons) and
automatically added
to the desktop for
easy access the next
time an MSN connec-
tion is established.

Newsreader Software

There are newsreaders for every computer platform, even those based on UNIX and DOS. The trouble with these types of newsreaders is that they are text-oriented and you must learn arcane commands to navigate through the list of newsgroups and the articles in each newsgroup. Macintosh and Windows newsreaders permit the novice to navigate the Usenet newsgroup maze with point-and-click ease from within the user's familiar graphical desktop environment. As noted, you can also read newsgroups with your Web browser or by using your America Online, CompuServe, or MSN software.

Newsreaders are also packaged differently. Some are stand-alone clients, such as Forte's Free Agent and Olav Tollefsen's News for Windows NT. Others come as separate applications within Internet suites, such as Delrina Corporation's Cyberjack for Windows 95—packaged with Delrina Comm Suite 95—and Quarterdeck Corporation's Message Center—included with InternetSuite for Windows. Still others are packaged as fully integrated Internet applications, such as Attachmate Corporation's Emissary (Fig. 10.9). Finally, Web browsers such as Mosaic and Netscape Navigator can include an integral newsreader.

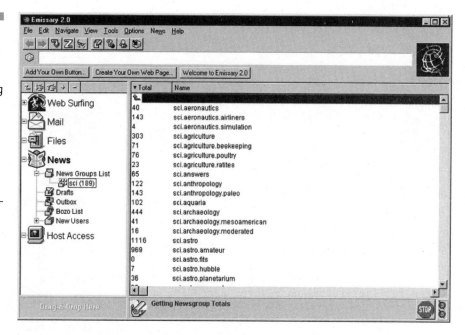

Figure 10.9
Attachmate's Emissary is a fully integrated Internet application, providing access to your PC, as well as the Web, e-mail, news, and hosts (for Telnet and FTP sessions)—all from a single Windows interface.

Stand-alone newsreaders are available from numerous vendors as stand-alone freeware and shareware. Many of these programs are more adept at helping you manage the hundreds—or even thousands—of messages that often appear in a single newsgroup and in helping you navigate through the various threads. They are also better at helping you download new messages, stitch together multipart file attachments, and apply filters so you don't get overwhelmed with information and possibly run out of disk space. Once you join a newsgroup, it is generally much easier to participate and stay organized if you use a stand-alone newsreader.

Many of these newsreaders are available from various Web and FTP sites on the Internet. They are usually small, compared to other programs, and download quickly into your computer. After trying out various newsreaders for awhile you will develop a preference for one or two products and stick with them. You can then register your favorite newsreader with the vendor—paying a small shareware fee, if any—and stay abreast of new product developments. Often, the next version of the software is free when you register.

A good newsreader helps you do three things: sort through the thousands of newsgroups available over Usenet, view message titles in a convenient list format, and read/reply to individual messages. Some newsreaders, such as Free Agent, also show you how messages and their replies relate to each other—even marking or automatically retrieving articles that belong to the same thread so you can more easily follow the various discussions.

To help you choose which newsgroups you want to subscribe to, the newsreader should provide a browse feature, allowing you to simply scroll through a list of available groups, and mark them for later retrieval. Navigation is further aided by a search capability that lets you find groups by key words in newsgroup titles. For example, searching on the word *clinton* turns up the following newsgroups:

alt.current-events.clinton.whitewater

alt.impeach.clinton

alt.politics.clinton

alt.president.clinton

And if you find a newsgroup you want to join, you should be able add its title to a subscribe list that appears every time you launch your newsreader. Not only that, but you should have the option of only retrieving new articles that have been posted since you were last online.

Once you're in a newsgroup, your newsreader displays the messages in a list. Since newsgroups are essentially discussion groups, there may be several replies associated with each article. The newsreader groups articles and replies together into a thread for display. For example, you may see the original subject of a message listed on a single line, with subsequent replies indented under it. You can open the first message and then click your way through the replies (Fig. 10.10).

Some newsreaders also offer a quote feature, which lets you copy text from an article and paste it into your reply, using a different color and/or indentation scheme (i.e., double angle brackets) to separate it from your response. This not only provides a convenient way to organize your response, it helps other readers understand the context of your reply.

As noted, many newsgroups are known for providing articles with binary attachments, such as images, video clips, and sound files. However, the newsgroup system was not originally designed for multimedia. Therefore, these types of files must be coded when posted, and then decoded when they are downloaded to your computer. In addition to UUE (UNIX-to-UNIX Encoding) and MIME (Multipurpose Internet Mail Extensions), you may encounter other coding schemes. A good newsreader supports all of them.

Large binary files often appear as several message attachments, which your newsreader should automatically stitch together as you download and split apart as you upload. And because downloading and uploading

Figure 10.10

In Free Agent, the
alt.illuminati news-
group is selected. To
the right, a particular
article in that news-
group is selected for
viewing (e.g., Evolu-
tionists Disproven).
All the replies associ-
ated with that article
are listed under it.
The text of the origi-
nal article is displayed
in a frame at the bot-
tom of the window.

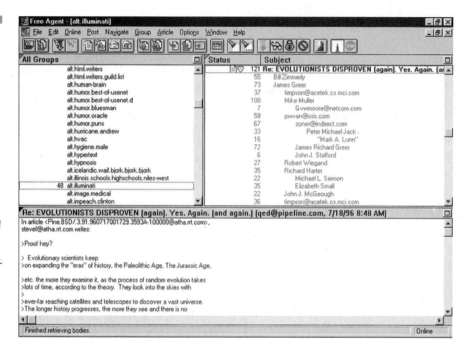

large binaries takes time, the newsreader should be able to perform these tasks in the background while you read or write messages in the foreground.

Some newsreaders require you to get very involved in downloading multipart binary attachments: you must highlight all parts of the file, mark them to be retrieved, and then save the files. Others are highly automated, only requiring that you double-click on the first part. The rest is automatic: the individual parts are retrieved in order, decoded, and then displayed in a viewing window. If the file type is not supported, some newsreaders can even be configured to launch an external viewer.

As you gain experience with newsgroups, you'll want a newsreader that can download unread messages quickly, so you can read them offline at a more convenient time and save money by minimizing your Internet access connection time, if you subscribe to a usage-based service. If these are important concerns, you'll want a newsreader that gives you the option to block certain downloads—based on such criteria as keyword, address, and whether the message includes a binary attachment. This type of filtering can help prevent you from getting overwhelmed with information and save space on your hard disk. Some newsreaders, such as Free Agent, let you choose what header fields you want to save: no fields (body only); newsgroup, subject, author, and date; or simply, all fields (Fig. 10.11).

Figure 10.11

With the Free Agent newsreader, you can specify what parts of the article you want to save to disk and append subsequent articles to an existing file.

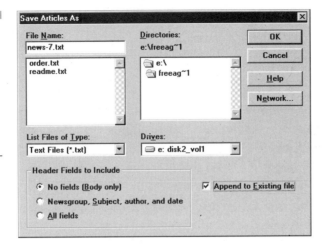

There is also a checkbox option that allows you to save the selected file by appending it to an existing file.

Features

Like other types of software, newsreaders differ in terms of performance and features. After trying out a few newsreaders to see how well they perform, you can focus in on comparing features to make a final selection. What follows is a list of features you may want to consider.

- *Address Book.* Allows you to keep track of e-mail addresses and other contact information.

- *Agents.* Automates routine tasks such as clean up directories, delete temporary files, check for new mail, check for new news, send automatically generated messages. Agents can be scheduled to run by time of day and day of week.

- *Article alert.* Allows you to configure the newsreader to periodically check for new articles in subscribed newsgroups and receive a visual or audible notification that articles of interest have been posted.

- *Associative subject threading.* Allows you to group together like subjects for convenient article viewing or downloading.

- *Context-sensitive help.* Provides help for the function or feature you happen to be using at the time you click on the help button.

- *Copy-self.* Allows you to send a copy of an e-mail message or article to yourself so you can have a permanent record of the item, including all the header information.

- *Courtesy copy.* Allows you to specify an e-mail address to receive a courtesy copy (cc:) of an e-mail message or article.

- *Cross-post management.* Reading a cross-posted article in one group marks it read in all other groups, so you don't end up wasting time reading duplicate articles.

- *Customizable toolbar.* Allows you to choose which commands appear on the newsreader's toolbar.

- *Drag and drop.* You can click on an object, such as a file, and drag it into a mail or news message as an attachment.

- *Duplicate file handling.* This feature prevents new downloaded files from overwriting existing files with the same name. Conflicts are resolved by automatically renaming the second file during the download so that it arrives with another name.

- *File extension translation.* Upon downloading files, this feature automatically changes file extensions from four-character UNIX extensions such as .jpeg or .html to three-character DOS extensions such as .jpg or .htm.

- *Filtering.* You can choose to have only a subset of the messages displayed, such as only unread articles.

- *Flat mode support.* A feature that allows you to sort all the articles in different ways—to see all articles from a particular author or in a particular order—without regard to the thread to which they belong.

- *Folders.* Allows you to store downloaded articles in any number of folders. You can name the folders in ways that help keep your articles organized for later retrieval.

- *Frames.* Displays separate frames within the main window: one each for the newsgroup list, article list, and message body. The individual frames can be sized within the main window.

- *Image viewer.* Allows you to view common image formats such as GIF and JPEG from within the newsreader.

- *International Character Support.* Displays characters from various languages in messages using the standard MIME format.

- *Kill list.* You can choose to have articles ignored automatically—so that they are not displayed—based on such criteria as subject or author. (Also called a *Bozo List.*)

▪ *Lock.*   Allows you to lock (and unlock) articles to prevent access or deletion.

▪ *MAPI support.*   Via support for Microsoft's Mail Application Program Interface (MAPI), you can send messages with attachments composed in other MAPI-compliant applications such as word processors and spreadsheets.

▪ *Message purging.*   Allows you to purge articles from your hard disk, freeing up more space. Articles can be purged when you receive new article headings from the same newsgroup; old articles can be purged automatically upon closing the newsreader; or articles can be purged on demand by subscribed newsgroup, selected newsgroups, or all newsgroups.

▪ *MIME attachments.*   Allows you to encode and decode binary attachments using the e-mail standard MIME format.

▪ *Multiple signatures.*   Allows you to create a separate signature for every newsgroup you post to.

▪ *Multiple windows.*   Allows you to open multiple windows, permitting you to do multiple tasks at the same time. For example, you can view multiple newsgroups in separate windows, switching between them to search and download articles.

▪ *Notification.*   The newsreader can be configured to respond to various events, such as the arrival of new e-mail or completion of a file download, by playing sounds or displaying messages.

▪ *Offline e-mail support.*   Allows you to retrieve e-mail messages from a POP or SMTP server and read them offline at a more convenient time or to minimize connection charges.

▪ *Quote support.*   Allows you to copy relevant passages of an article and paste them into your reply so that your response can be read in context.

▪ *Read/write newsrc files.*   Enables the use of multiple newsreaders—one at home and the other at work, for example. This feature keeps track of which articles have been read by both programs.

▪ *Receive e-mail.*   Allows you to receive e-mail from POP and SMTP mail servers.

▪ *Rules-based e-mail.*   Allows you to have incoming e-mail sent to specific folders based on the presence of key words or phrases in the messages.

■ *Scalable images.* This feature allows you to scale downloaded images up or down for viewing or make them as large as the window will allow.

■ *Search.* Allows you to find articles in a newsgroup by subject or key word.

■ *Send e-mail.* Allows you to respond to just the author of an article via his or her e-mail address instead of posting a response to the newsgroup.

■ *Sense typing.* As you type in a key word for a newsgroup, the newsreader shows a list of matching groups (Fig. 10.12). As more letters are typed, the list is narrowed further until the desired newsgroup name is displayed.

■ *Signature.* Allows you to create a standard signature or sign-off that is automatically placed at the end of your articles.

■ *Sorting.* You can have articles and e-mail messages displayed in a preferred order, such as by author, date, or file size.

■ *Spell Checker.* Allows you to spell check words against a built-in dictionary before posting your articles. You can customize the dictionary by adding your own words.

■ *SSL support.* Safeguards communication between clients and news servers via the Secure Sockets Layer (SSL) protocol, which provides encryption and authentication.

Figure 10.12
The sense-typing feature of Trumpet Newsreader. As letters are typed, appropriate newsgroup names are displayed. This feature helps you find newsgroups with arcane names.

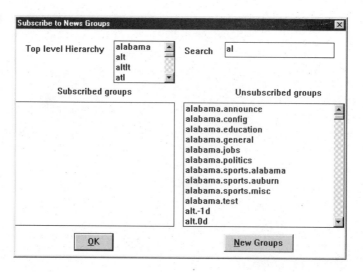

- *Stitching.* Automatically puts together multipart binary attachments when they are downloaded, even if they were not posted in the proper order.

- *Tagging.* Lets you tag articles so that you can return to them later to save, print, or reply to them.

- *Templates.* Among other things, templates allow you to create a header that automatically plugs in dates and names when you quote other messages. This helps everyone understand the context of a message thread. You can also decide how the subject line will appear when you post multipart binary messages, configuring the newsreader to display the number and sequence of files, for example.

- *Themes feature.* Allows you to create custom collections of macro buttons and bookmark pages and quickly switch between them. This is useful for creating collections of functions you need for work or entertainment or when you're traveling with your portable computer.

- *Thread support.* Allows you to easily follow discussions by having all replies appear with their respective articles. (See also the previous description of flat mode support.)

- *URL launcher.* You can automatically open your Web browser or FTP program by clicking on a hypertext link within an article. The hypertext link not only opens these applications automatically, but brings you to the location specified by the link.

- *Watch list.* You can choose to retrieve articles automatically based on such criteria as subject or author.

Online Resources

The following table provides links to the Web pages or FTP sites of the major newsreader providers. From these links, you can download working or evaluation copies of newsreader software. You can also obtain other useful information such as the platforms supported, system requirements, product features, installation instructions, and troubleshooting advice. Since the technology is moving rapidly, you may want to consult these locations periodically for the latest developments.

Developer	Product	Web page or FTP site
Attachmate Corp.	Emissary	http://www.attachmate.com/
Scott Baker	SBNews for Windows 95	http://www.primenet.com/~smbaker/sbnews/sbnews.html
Bob Boonstra	Value-Added NewsWatcher	http://www.ultranet.com/~boonstra/
Roger Brown	InterNews	http://dartmouth.edu/~moonrise
Delrina Corp.	CyberJack	http://www.delrina.com
Forte	Free Agent, Agent	http://www.forteinc.com
Intel Corp.	Smart Newsreader for Windows 95	http://www.intel.com/iaweb/aplets/newsread.htm
MicroPlanet	MicroPlanet Gravity	http://www.microplanet.com
Microsoft Corp.	Internet Explorer	http://www/microsoft.com
Netscape Communications Corp.	Netscape News	http://www.netscape.com
John Norstad	NetWatcher	ftp://ftp.acns.nwu.edu/pub/newswatcher/
Peter Speck	Nuntius	http://helpdesk.uvic.ca/how-to/support/mac/nuntius.html
SW15 Software Ltd.	NewsHopper	http://www.demon.co.uk/sw15/
Quarterdeck Corp.	Message Center	http://www.quarterdeck.com
Mark Riordan	WinVN	http://www.ksc.nasa.gov/software/winvn/winvn.html
Olav Tollefsen	News for Windows 95	http://193.213.32.10:80/NTNews/
Trumpet Software International	Newsreader for Windows	http://www.trumpet.com/

NOTE: This information, as well as updates, can be found at http://www.ddx.com/mgh.shtml.

Conclusion

To the novice opening a newsgroup for the first time, the sight of a seemingly endless list of article postings might seem like complete anarchy. Initially, you may find it difficult to make sense of it all. You may even find the topics of some newsgroups and the language used by authors to be vitriolic and offensive. That's okay—you do not have to participate in such discussions and you can even configure your newsreader to ignore these groups so they are not even displayed in a list. If you've ever heard the phrase *marketplace of ideas,* newsgroups are its manifestation.

Fortunately, newsreaders can help you find your way around this marketplace of ideas. You can subscribe to newsgroups that discuss only topics that are relevant to your personal and professional life, such as your hobby and recreational interests, job-related interests, sports and entertainment interests, or investment interests. You also have the freedom to explore newsgroups that discuss topics you would never have considered before. With 15,000 to 20,000 newsgroups in operation, there's always something new to discover that will engage you more than television ever can.

How can you tell when you are *engaged?* Quite simply: if you suddenly look up from your computer and are astonished to find that three hours have already gone by, you're engaged!

APPENDIX A

REMOTE PRINTING COVERAGE LIST

In numerical order by country code

CANADA AND THE UNITED STATES (+1)

+1-205 Birmingham, Alabama (Metropolitan)

+1-209 Fresno, Clovis

+1-212 Manhattan

+1-310 La Habra, Pico Rivera, Whittier

+1-313

+1-317 Battle Ground, Brookston, Buck Creek, Clarks Hill, Lafayette/West Lafayette (Purdue University), Mulberry, Otterbein, Romney, West Point

+1-360 Washington (Southern)

+1-412 Carnegie Mellon University (Pittsburgh)

+1-415 Conta, Belvedere, San Francisco 1—3, Sausalito

+1-416 SERVICE SUSPENDED UNTIL FURTHER NOTICE Metro Toronto, University of Toronto, St. George Campus and Scarborough Campus, Royal Conservatory of Music, Victoria College, Aerospace Studies, Mt. Sinai Hospital, Mt. Michael's Hospital, and College Hospital for Sick Children, Sunnybrook Hospital, East General Hospital, Toronto Hospital, Women's College Hospital, Newman Centre, Centre for Health Promotion, St. Joseph's College

+1-503 Oregon (complete coverage)

+1-510 Antioch, Clayton, Concord 1, Danville, East Bay 1—4, Lafayette, Martinez, Moraga, Orinda, Pittsburg, Pittsburg West, Richmond, Walnut Creek

+1-514 Montreal

+1-516 Floral Park, Cedarhurst, Valley Stream, Great Neck, Manhassett, Port Washington

+1-541 Oregon (complete coverage)

+1-613 Almonte, Alymer, Bourget, Buckingham, Carlton Place, Carp, Casselman, Chelsea, Clarence Creek, Constance Bay, Crysler, Cumberland, Embrun, Gatineau, Gloucester, Jockvale, Kanata Stittsville, Kemptville,

Low, Luskville, Manotick, Merrickville, Metcalfe, Navan, North Gower, Orleans, Osgoode, Ottawa Hull, Pakenham, Perkins, Plantagenet, Quyon, Richmond, Rockland, Russel, St. Pierre de Wakefield, Thurso, Wakefield

+1-617 Arlington, Belmont, Cambridge, Lexington, Medford, Somerville, Winchester

+1-707 Benicia

+1-714 Brea, UC Irvine

+1-718 Cambridge Computer Associates, N.Y., Queens, Brooklyn, and the Bronx, N.Y., Tinkelman Enterprises and Staten Island, N.Y.

+1-800 Toll-free

+1-813 Tampa

+1-818 Arcadia, Azusa, Covina 1—2, El Monte, La Puente, Monrovia, San Gabriel Canyon, Sierra Madre

+1-819 Alymer, Buckingham, Chelsea, Gatineau, Low, Luskville, Perkins, Quyon, St. Pierre de Wakefield, Thurso, Wakefield

+1-905 SERVICE SUSPENDED UNTIL FURTHER NOTICE University of Toronto, Erindale Campus

+1-909 Chino, Claremont, Diamond Bar, Ontario, Pomona, Upland 1—2

+1-917 New York City, N.Y.

+1-919 Chapel and Carrboro, Mebane, Hillsborough, Pittsboro, RTP, RDU Airport, University of North Carolina at Chapel Hill

GREECE (+30)

Athens (+30 1)
Thessaloniki (+30 31)
Iraklio (+30 81)

PORTUGAL (+351)

Lisbon (+351—1)

CROATIA (+385)

Zagreb (+385—1)

ITALY (+39)

Pisa (+39—50 select numbers only)

UNITED KINGDOM, IOM, AND THE CHANNEL ISLANDS (+44–complete coverage)

DENMARK (+45)

Lyngby (+45-45)

SWEDEN (+46–complete coverage)

AUSTRALIA (+61)

Australia-freecall (+61-1-800)
Canberra, Melbourne, Trafalgar, Traralgon, Kilmore, Emerald, Healesville, Mornington, Rosebud, Cranbourne, Pakenham, Morwell, Lara, Bacchus Marsh, Balliang, Gismore, Romsey, Sydney, Darwin

NEW ZEALAND (+64)

Wellington (+64—4)
Hamilton (+64—7823, 7824, 7825, 7827, 7829, 783, 784, 785)

KOREA (+82)

Seoul (+82—2)

HONG KONG (+852)

CHINA (+86)

Sichuan Province—Chengdu City (+86—28)

TAIWAN (+886)

Tauyuan county (+886—3)

APPENDIX B

GUIDELINES FOR USENET GROUP CREATION

Archive-name: creating-newsgroups/part1

Original-author: woods@ncar.ucar.edu (Greg Woods)

Comment: enhanced & edited until 5/93 by spaf@cs.purdue.edu (Gene Spafford)

Last-change: 17 Feb 1994 by tale@uunet.uu.net

REQUIREMENTS FOR GROUP CREATION

These are guidelines that have been generally agreed upon across USENET as appropriate for following in the creating of new newsgroups in the "standard" USENET newsgroup hierarchy. They are NOT intended as guidelines for setting USENET policy other than group creations, and they are not intended to apply to "alternate" or local news hierarchies. The part of the namespace affected is comp, news, sci, misc, soc, talk, rec, which are the most widely-distributed areas of the USENET hierarchy.

Any group creation request which follows these guidelines to a successful result should be honored, and any request which fails to follow these procedures or to obtain a successful result from doing so should be dropped, except under extraordinary circumstances. The reason these are called guidelines and not absolute rules is that it is not possible to predict in advance what "extraordinary circumstances" are or how they might arise.

It should be pointed out here that, as always, the decision whether or not to create a newsgroup on a given machine rests with the administrator of that machine. These guidelines are intended merely as an aid in making those decisions.

THE DISCUSSION

1) A request for discussion on creation of a new newsgroup should be posted to news.announce.newgroups, and also to any other groups or mailing lists at all related to the proposed topic if desired. The group is moderated, and the Followup-to: header will be set so that the actual discussion takes place only in news.groups. Users on sites which have difficulty posting to moderated groups may mail submissions intended for news.announce.newgroups to newgroups@uunet.uu.net.

The article should be cross-posted among the newsgroups, including news.announce.newgroups, rather than posted as separate articles. Note that standard behaviour for posting software is to not present the articles in any groups when cross-posted to a moderated group; the moderator will handle that for you.

2) The name and charter of the proposed group and whether it will be moderated or unmoderated (and if the former, who the moderator(s) will be) should be determined during the discussion period. If there is no general agreement on these points among the proponents of a new group at the end of 30 days of discussion, the discussion should be taken offline (into mail instead of news.groups) and the proponents should iron out the details among themselves. Once that is done, a new, more specific proposal may be made, going back to step 1) above.

3) Group advocates seeking help in choosing a name to suit the proposed charter, or looking for any other guidance in the creation procedure, can send a message to group-advice@uunet.uu.net; a few seasoned news administrators are available through this address.

THE VOTE

The Usenet Volunteer Votetakers (UVV) are a group of neutral third-party vote-takers who currently handle vote gathering and counting for all newsgroup proposals. Ron Dippold <rdippold@qualcomm.com> co-ordinates this group. Contact him to arrange the handling of the vote. The mechanics of vote will be handled in accord with the paragraphs below.

1) AFTER the discussion period, if it has been determined that a new group is really desired, a name and charter are agreed upon, and it has been determined whether news.announce.newgroups and any other groups or mailing lists that the original request for discussion might have been posted to. There should be minimal delay between the end of the discussion period and the issuing of a call for votes. The call for votes should include clear instructions for how to cast a vote. It must be as clearly explained and as easy to do to cast a vote for creation as against it, and vice versa. It is explicitly permitted to set up two separate addresses to mail yes and no votes to provided that they are on the same machine, to set up an address different than that the article was posted from to mail votes to, or to just accept replies to the call for votes article, as long as it is clearly and explicitly stated in the call for votes article how to cast a vote. If two addresses are used for a vote, the reply address must process and accept both yes and no votes OR reject them both.

2) The voting period should last for at least 21 days and no more than 31 days, no matter what the preliminary results of the vote are. The exact date that the voting period will end should be stated in the call for votes. Only votes that arrive on the vote-taker's machine prior to this date will be counted.

3) A couple of repeats of the call for votes may be posted during the vote, provided that they contain similar clear, unbiased instructions for casting a vote as the original and provided that it is really a repeat of the call for votes on the SAME proposal (see #5 below). Partial vote results should NOT be included; only a statement of the specific new group proposal, that a vote is in progress on it, and how to cast a vote. It is permitted to post a "mass acknowledgment" in which all the names of those from whom votes have been received are posted, as long as no indication is made of which way anybody voted until the voting period is officially over.

4) ONLY votes MAILED to the vote-taker will count. Votes posted to the Net for any reason (including inability to get mail to the vote-taker) and proxy votes (such as having a mailing list maintainer claim a vote for each member of the list) will not be counted.

5) Votes may not be transferred to other, similar proposals. A vote shall count only for the EXACT proposal that it is a response to. In particular, a vote for or against a newsgroup under one name shall NOT be counted as a vote for or against a newsgroup with a different name or charter, a different moderated/unmoderated status or (if moderated) a different moderator or set of moderators.

6) Votes MUST be explicit; they should be of the form "I vote for the group foo.bar as proposed" or "I vote against the group foo.bar as proposed". The wording doesn't have to be exact, it just needs to be unambiguous. In particular, statements of the form "I would vote for this group if..." should be considered comments only and not counted as votes.

7) A vote should be run only for a single group proposal. Attempts to create multiple groups should be handled by running multiple parallel votes rather than one vote to create all of the groups.

THE RESULT

1) At the completion of the voting period, the vote-taker must post the vote tally and the E-mail addresses and (if available) names of the voters received to news.announce.newgroups and any other groups or mailing lists to which the original call for votes was posted. The tally should include a statement of which way each voter voted so that the results can be verified.

2) AFTER the vote result is posted, there will be a 5 day waiting period, beginning when the voting results actually appear in news.announce. newgroups, during which the Net will have a chance to correct any errors in the voter list or the voting procedure.

3) AFTER the waiting period, and if there were no serious objections that might invalidate the vote, and if 100 more valid YES/create votes are received than NO/don't create AND at least 2/3 of the total number of valid votes received are in favor of creation, a newgroup control message may be sent out. If the 100 vote margin or 2/3 percentage is not met, the group should not be created.

4) The newgroup message will be sent by the news.announce.new-groups moderator at the end of the waiting period of a successful vote. If the new group is moderated, the vote-taker should send a message during the waiting period to David C. Lawrence.

5) A proposal which has failed under point (3) above should not again be brought up for discussion until at least six months have passed from the close of the vote. This limitation does not apply to proposals which never went to vote.

GLOSSARY

absolute URL A reference to a document or service located on another server on the Internet. (See also **relative URL**.)

ACTA America's Carriers Telecommunication Association (ACTA), a trade association of 130 competitive, long-distance carriers, which filed a protest with the FCC concerning the increasing use of the Internet for telephone calls.

ADSL Asymmetric Digital Subscriber Line (ADSL) is an affordable local loop upgrade technology that allows telephone companies to offer transmission rates of more than 6 Mbps over existing twisted-pair copper wiring—1.544 to 6.144 Mbps downstream (to the customer) and 16 to 640 Kbps upstream (to the telephone company). This is enough bandwidth to support multimedia—video, audio, graphics, and text—to homes and businesses. This technology will also make Web applications work almost as fast as the applications that run on your computer.

aliases In the context of IRC chat software, aliases are custom commands that users can define for their own use but which are implemented on an IRC server in their standard form. They are simply short, easy-to-remember substitutes for longer text-based commands.

analog The traditional format for telephone and cable television transmission, in which sound and video are converted into continuous electrical impulses that can be carried by radio waves. Increasingly, the analog format is being replaced by a digital one. (See also **sampling**.)

anchor In HTML parlance, an anchor defines the position of an inline image or text which is a hypertext link. Clicking on the anchor brings the user to another file or service to which the link points.

Application Programming Interface (API) A language and message format used by an application program to communicate with another program, allowing data to be transferred and used between them. APIs are usually implemented by writing function calls. Examples of APIs are the calls made by an application program to such programs as an operating system, messaging system, or database management system.

ASCII American Standard Code for Information Interchange offers a standard for representing computer characters. There are 128 standard ASCII character codes, each of which can be represented by a 7-digit binary number: 0000000 through 1111111. ASCII files often have the .txt file extension.

automatic gain control   A feature of some sound cards that boosts the microphone level automatically when the user speaks into it and reduces the level when the user is not speaking. This cuts down on ambient background noise that makes it difficult for both parties to hear each other during an Internet telephone conversation.

BACP   The Bandwidth Allocation Control Protocol (BACP) is an outgrowth of the Point-to-Point Protocol Multilink specification, which lets users call up additional bandwidth on an ISDN line via a router or terminal adapter. This gives clients the ability to dynamically add 56/64 Kbps channels as needed to support the application and then drop them when the extra bandwidth is no longer needed.

bandwidth   Refers to the capacity of data that can be transmitted across wireline and wireless networks. Bandwidth size affects the speed with which data can be transmitted.

BBS   A bulletin board system, usually offering dial-up access to local users. However, BBSs are accessible from anywhere over the public telephone network. BBSs also can be set up for corporate use, which can be reached via their private data networks, as well as the public telephone network.

BinHex   In the Macintosh world, documents are coded into BinHex for transmission over the network. Upon delivery, the document is decoded to its original form so it can be understood by the application. BinHex includes error checking and compression.

broadcast   One-way transmission of voice, video, or text to many recipients through a specially equipped server that replicates the data stream and sends it out to authorized users often over different paths through the network.

browser   A computer application that is used to view documents on the World Wide Web. The browser renders documents that are coded in the HyperText Markup Language (HTML). Although browsers cannot edit the documents they display, browsers can be used to save the documents to disk where they can be edited by word processors.

byte   A common unit of computer file size, with one byte usually consisting of eight bits. A bit consists of one character whose value can be 1 (on) or 0 (off).

cache   A location in memory where data is temporarily stored for easy retrieval. Most Web browsers make use of cache, allowing users to view previously loaded HTML documents without having to return to the Web to retrieve them again.

Cellular Digital Packet Data (CDPD) An open specification for a wireless data network developed by AT&T Wireless Services, IBM, and a consortium of U.S. cellular providers. CDPD allows data to be transmitted wirelessly over the cellular voice network. Because it transmits information using the Internet Protocol (IP), CDPD is the wireless equivalent of the Internet. This relationship makes it the most suitable solution for mobile applications requiring a wireless extension to the Internet and corporate intranets.

CERN The European Laboratory for Particle Physics where HTML was developed in the late 1980s to provide researchers with a means to access and display documents stored on servers anywhere on the Internet.

CGI Common Gateway Interface, a method of interfacing Web servers with programs, allowing them to process information based on forms and other user inputs such as hypertext links.

cgi-bin directory A directory in which CGI scripts are held, such as those that process information entered into HTML forms posted on the Web.

chat An interactive mode of communication that entails the typing of messages between two people or between various members of a group, all of whom are logged on to the network at the same time.

chat rooms A server or channel in which users chat with each other in real time by issuing text commands. More sophisticated chat rooms are graphical in nature, allowing users to move avatars and other objects through an online shopping mall or fantasy world.

circuit-switched cellular data A method of transmitting data over a cellular voice channel in a manner similar to that of a conventional telephone connection. Unlike CDPD connectivity, which is established without a dial-up procedure, circuit-switched users must dial an Internet access provider to establish a connection. Once connected, the user occupies that circuit until the connection is terminated.

client A program that requests information from a server. Web browsers such as Netscape and Mosaic are client applications because they request documents from servers on the World Wide Web, Gopher documents from Gopher servers, files from FTP servers, and articles from USENET news servers.

Client-to-Client Protocol A protocol that allows users to retrieve information about other users on a chat channel, such as their real name, e-mail address, and the kind of chat software they are using.

client-server A networkcentric method of computing in which processing duties are distributed among workstations and servers in a network to improve network speed, efficiency, and security. The Internet is an example of a wide area client-server network.

clipping A condition in which there is too much information loss during transmission, resulting in decompressed voice at the receiving end getting cut off in midsentence. (See also **dropouts.**)

command line interface The traditional method of issuing commands and manipulating data on the computer via text commands versus a graphical user interface (GUI) where all commands are issued by picking commands off pull-down menus, dragging and dropping objects, and opening windows and dialog boxes.

compression A software or hardware technique that shrinks the computer data before it is transmitted over the Internet. At the other end, the data is decompressed to its original size. Compression is particularly important for the transmission of voice and video, which in their uncompressed forms consist of a huge number of bits, making it slow and difficult to transmit.

Cyberspace A casual term used to describe the universe of internetworked computers that comprise the Internet.

daemon Disk and execution monitor, a program that is not explicitly invoked, but which waits for some condition(s) to occur before springing into action. This saves end users from having to understand the idiosyncrasies of a process, such as printing a file or sending e-mail. Instead, the daemon decides how to handle the process.

database A collection of files, such as HTML documents on a Web server.

decoder An application that decompresses audio or video files to their original size so they can be played back. The decoder usually has controls that allow you to do such things as start/stop or pause the recording, adjust the volume, and specify audio/video quality. (See also **encoder.**)

dedicated line A full-time, high-speed connection to the Web server. This is the most expensive type of Internet connection, the cost of which is usually justified by a high volume of traffic.

Direct Client to Client A protocol that allows two parties to chat with each other directly, rather than through the IRC network, using only e-mail addresses. DCC also allows files to be sent and received between users. Users have the choice of whether to accept or reject such transactions.

dither   A method of creating the illusion of a color by placing dots of other colors very close together. A color is dithered when the display adapter does not support that color.

document source   The ASCII version of a document being rendered by the browser, showing all HTML tags used to code the document. Many browsers include the capability to view the document source through a pop-up window.

domain name server   A computer that matches domain names to numeric addresses, making them easier to remember. All Web servers have this capability.

download   The transfer of a file from a remote system to a local system. For example, documents rendered by a browser can be downloaded from the remote Web server to the local client computer via the save as command (or something similar) accessed from File on the menu bar.

dropouts   A condition in which packets of voice are lost, resulting in a missing segment of speech at the receiving end of a conversation. This is also called *clipping*.

DSVD   Digital Simultaneous Voice and Data is a technology used in newer 28.8-Kbps modems that allows voice as well as data to be transmitted over the line simultaneously. DSVD modems have audio input and output ports and handle all the voice compression and decompression, as well as the transmission. Since the PC never has the opportunity to manipulate voice, this function is totally independent of any application on the computer.

Dynamic Host Configuration Protocol (DHCP)   A method of issuing IP addresses developed by the Internet Engineering Task Force (IETF). From a pool of IP addresses, a DHCP server doles them out to users as they establish Internet connections. When they log off the Net, the IP addresses become available to other users.

dynamic IP address   An IP address that is assigned by an Internet service provider each time a user dials into the server. The proliferation of TCP/IP-based networks, coupled with the growing demand for Internet addresses, makes it necessary to conserve IP addresses. Issuing IP addresses on a dynamic basis provides a way to recycle this finite resource. (See also **static IP address.**)

e-mail   An abbreviation for *electronic mail,* which refers to the ability of a system to send and receive messages over a network.

encoder   An application that enables you to record audio or video files. The encoder compresses the files to minimize the byte count. In compressed form, the files can be stored on disk or sent over the Internet more efficiently. The encoder includes controls that allow you to set the sampling rate and adjust the recording volume. (See also **decoder.**)

encryption   A process whereby data is scrambled to protect it from unauthorized access. At the opposite end, the data is decrypted to its original form so it can be read. A popular and effective encryption program is Phil Zimmerman's PGP (Pretty Good Privacy).

external image   An image that is specifically requested by activating a hypertext link. (See also **inline image.**)

FAQ   Short for *frequently asked questions.* A FAQ file contains the answers to commonly asked questions to expedite the handling of routine inquiries, while freeing up disk space on a news server or bulletin board system (BBS).

Federal Communications Commission   The federal agency that is responsible for carrying out telecommunications policy in the United States. This includes promulgating regulations to implement the Telecommunications Act of 1996.

feedback   A situation that results when a microphone picks up sound from nearby speakers. In the context of telephone conversations over the Internet, feedback can cause the user at the other end of the connection to hear his or her own voice as an echo. Feedback can be eliminated by moving the microphone and speakers farther apart or by using a headset instead of speakers.

Finger   A command that is used to display information about a user on a remote system, such as when he or she last logged on. To use Finger, you must know the user name and host name of that individual. In addition, the remote system must be configured to accept remote Finger requests.

firewall   Software and/or hardware that provides security to a system connected to the Internet, preventing intruders from gaining access to private information. A firewall enforces an organization's security policies by limiting what kind of traffic that can pass in each direction.

FTP   File Transfer Protocol, used to transfer files between systems on the Internet. A special use of FTP is *anonymous FTP,* which provides public access for file transfers to a computer system that would otherwise require special authorization, such as a username and password.

full duplex   A mode of two-way communication that allows each party to speak simultaneously without having to wait for the link to be relinquished at the other end, as in the half-duplex mode of communication.

gateway   A program that translates information going between different computer systems or different applications.

GIF   Graphics Interchange Format, a type of file format widely used on the Internet, especially in HTML documents accessible on the Web. This file format was originally developed by CompuServe.

Gopher   A menu-based system, usually text only, for finding information on the Internet. Unlike HTML documents on the Web, Gopher documents do not contain hypertext links.

Gopherspace   A casual term used to describe the worldwide network of Gopher systems.

GPS   The Global Positioning System (GPS) is a network of 24 Navstar satellites in orbit at 11,000 miles above the earth. The GPS can be used to pinpoint locations anywhere on or above the earth to within a few centimeters. The system is used for defense systems and navigation as well as mapping and telelocator applications.

GSM   Global System for Mobile communications is a worldwide standard for digital cellular communications. It provides close to a 5:1 compression of raw audio with an acceptable loss of audio quality on decompression. Vendor-specific variants of GSM are also used for compressing voice for transmission over the Internet.

GUI   Graphical user interface. A graphical environment in which commands are issued by pointing and clicking items with a mouse. Data manipulation is accomplished by such methods as picking commands off pull-down menus, dragging and dropping objects, and opening windows and dialog boxes. (See also **command line interface.**)

half duplex   A mode of two-way communication that works much like a citizen's band (CB) radio, where one person can talk at a time and says *over,* indicating when he or she is finished talking.

HDML   The acronym for *Handheld Device Markup Language,* an open programming language derived from the HyperText Markup Language (HTML), the fundamental application programming language of the Internet. HDML is specifically designed for developing Internet applications that run on small wireless devices, such as the AT&T PocketNet Phone.

helper application   An external program that a Web browser runs whenever it encounters a certain type of file. For example, you can set up Netscape Navigator to run an audio player whenever it comes across a WAV file on the Web. (See also **plug-in.**)

home page   The first HTML file that is usually accessed at a particular WWW site. A home page typically includes a menu of hypertext links which are used by readers to access related files and/or other resources on the Web.

host   A computer, particularly one that is the source or destination of messages on a communications network. On the Internet, a host can be any type of computer with its own IP address.

hotlist   A selection of interesting URLs compiled by the user, providing an easy way to return to various resources on the Internet.

hotspot   A portion of an image that changes the mouse cursor, indicating the location of a hypertext link.

HTML   HyperText Markup Language, a coding specification that relies on tags to indicate how documents on the World Wide Web should be rendered by browsers. HTML documents are portable from one platform to another. HTML documents are essentially SGML documents with generic semantics that are appropriate for representing information from a wide range of applications. HTML has been in use on the World Wide Web since 1990. (See also **SGML.**)

HTML editor   A program that facilitates the creation of HTML documents. HTML editors are available as stand-alone products or as add-ins to word processors.

HTTP   HyperText Transfer Protocol, used for transferring hypertext documents, such as those used on the World Wide Web, across the Internet.

hyperlink   A shortened form of hypertext link.

hypertext   A method of interlinking collections of documents so that users can go from one topic to another as their interests or information needs change from moment to moment.

hypertext link   A connection between two documents (or resources) built from HTML anchor tags, an example of which is <A HREF="filename.html">...</A>.

IDDD   International Direct Dialing Designator, a reference to how users dial international calls without operator assistance. IDDD is used in the remote-printer application for specifying the fax number of an addressee when sending e-mail to a fax machine.

IETF   Internet Engineering Task Force, a volunteer group that researches and solves technical problems and makes recommendations to the Internet Architecture Board (IAB).

image map   A graphic image containing hypertext links. The user clicks on a specific area of the image to go to another document.

inking   A capability of some conferencing and collaborative computing software that allows users to view and mark up the same document or spreadsheet over the Internet.

inline image   An image that appears within the text of a document. When the document is retrieved, the appropriate image file is automatically retrieved as well. The browser displays the image as if it were part of the text. (See also **external image.**)

Internet   A worldwide network of computers offering many types of services, including the World Wide Web.

Internet appliance   A term that has come into vogue recently, referring to a computing device with limited onboard computing power that would serve only to provide access to the Internet. Until recently, it has been considered a desktop device. However, it could include handheld wireless devices for mobile users.

InterNIC   InterNIC (Internet Network Information Center) is a cooperative project between the National Science Foundation (NSF), Network Solutions, and AT&T. NSF supports the Internet with grants, funding, and research. Network Solutions sponsors Registration Services (domain name registration and IP network number assignment), Support Services (outreach, education, and information services for the Internet community), and Net Scout Services (Internet research and reporting). AT&T supports Directory and Database Services (white pages and publicly accessible databases).

intranet   A private corporate network that uses Internet Protocol, Web server technology, and all or part of the Internet network infrastructure for data transmission. Intranets typically are partitioned from the public Internet through software technology known as a firewall so that proprietary information can be made available only to those authorized to see it.

IP address   A set of numbers uniquely identifying a particular computer connected to the Internet (such as 127.127.127.11). Because IP addresses can be difficult to remember, Internet software uses the domain name system (DNS) to translate these numeric addresses into easier to remember names, such as iquest.com or whitehouse.gov. (See also **domain name server.**)

IRC　An acronym for Internet Relay Chat, a global standard for exchanging text messages in real time (as opposed to merely posting messages via e-mail or news) via specially configured servers on the Internet. The main IRC network often has tens of thousands of users online at any given time. Via the IRC network users can participate in scheduled or ongoing chat sessions that are organized around themes or topics.

ISDN　An acronym for Integrated Services Digital Network, which is a carrier-provided, switched digital service that can handle voice and data at the same time.

ISO　International Organization for Standardization, a worldwide organization that issues international standards in many different technical areas, such as SGML. (See also **SGML.**)

ISP　Internet Service Provider, a company that offers local Internet access on a subscription basis. Dial-up SLIP/PPP accounts that access the Internet via modem typically cost between $10 and $20 per month.

jitter　Jitter is the uncertainty in the arrival time of a packet or the variability of latency. (See also **latency.**)

JPEG　Joint Photographic Experts Group, the international standards body that developed the compression/decompression algorithms for photographic image files. This file format can be much more compressed than, say, GIF, but high compression causes some loss of detail. Such files use the .jpg file extension.

LAN　Short for local area network, on which groups of computers are connected within an office or building. Ethernet and Token Ring are the most popular LANs. To connect these LANs in a campus environment, FDDI (Fiber Distributed Data Interface) or ATM (Asynchronous Transfer Mode) technologies may be used. LANs can be interconnected over greater distances via private TCP/IP intranets or via carrier-provided lines and services, which also are referred to as WANs—wide area networks.

latency　Latency is the delay between the time data is transmitted and the time it is received at the destination.

mail-to-fax gateway　Server-based software that gives Internet users the ability to send faxes as they would e-mail. The gateway software can send faxes to any machine—on or off the Internet. Off-Net faxes are sent to the last possible server on the Internet before dialing the public network to complete the transmission. This is also called *remote printing.*

MAPI   Microsoft's MAPI (Mail Applications Programming Interface) specification enables you to easily send and receive messages with attached documents from other MAPI-compliant applications such as Microsoft Word and Excel.

MBONE   MBONE is short for *multicast backbone.* In operation since 1992, the MBONE is an overlaid subnetwork of the Internet that provides multimedia conferencing through IP multicast routing. IP multicast routing enhances the ability of distributed applications to run in real time over wide area IP networks by giving routers the responsibility for distributing and replicating the multicast data stream closest to their destinations, as opposed to duplicating packets for transmission to individual IP hosts (i.e., each end user's computer). This prevents network congestion and conserves processing resources. (See also **multicast** and **unicast.**)

META   An element used within the HEAD element to embed document meta-information not defined by other HTML elements. Such information can be extracted by servers/clients for use in identifying, indexing, and cataloging documents.

MIME   Multipurpose Internet Mail Extension, a standard used by Internet e-mail applications that enables users to send virtually any type of data across the Internet, including text, graphics, sound and video clips, and many other types of files.

modem   Short for *modulator-demodulator.* The device that lets a computer communicate over telephone lines by turning digital computer data into analog signals for transmission and back again.

Mosaic   A browser that provides access to documents on the World Wide Web. Mosaic is written and maintained at the National Center for Supercomputing Applications (NCSA) at the University of Illinois.

MPEG   Moving Pictures Experts Group, the international standards body that developed the compression/decompression algorithms for video files. Such files are identified with the .mpg file extension.

MUD   Multi-User Dungeons are chat systems that are used mostly for games. They typically include graphical components that can be manipulated within a three-dimensional fantasy world. A hundred or more players may be active in the same MUD at any given time.

multicast   A highly efficient method of information delivery in which routers on the Internet are configured to perform the replication of packets at the point closest to the end user (i.e., host IP address), instead of having the source duplicate packets for delivery to every end user wishing to receive the broadcast. This conserves bandwidth on the Internet and eases the processing burden of the server. (See also **MBONE** and **unicast.**)

NCSA   National Center for Supercomputer Applications at the University of Illinois, where the Mosaic Web browser was originally developed.

newsgroup   A collection of articles on a specific topic that are posted on a news server. There are approximately 4000 newsgroups accessible from the Web, covering a wide range of topics.

NHTML   Netscape HTML refers to the HTML-like tags recognized by Netscape Navigator, a popular browser offered by Netscape Communications. These HTML-like tags are proprietary and are not part of HTML 2.0 or HTML 3.0 specifications. Two examples are the CENTER and BLINK tags.

NNTP   An acronym for *Network News Transfer Protocol,* which runs on a news server, enabling users to subscribe to newsgroups, as well as post and receive articles.

object   Any graphical element that appears within an HTML document, including a hypertext link.

PCMCIA   Personal Computer Memory Card International Association, the developer of the form factor known as PCMCIA. Initially, this form factor (a card) was limited to memory cards that added more RAM to portable computers. Now several devices have been packaged into PCMCIA cards, including fax/modems, LAN adapters, hard drives, and specialized connectors for external devices such as CD-ROMs. PCMCIA is now known as PC Card, but the old name is still commonly used.

PCS   A new category of wireless network services, known as Personal Communications Services (PCS). Narrowband PCS will be used to provide such new services as advanced voice paging, two-way acknowledgment paging, and data services, while broadband PCS will be used to provide a variety of new mobile services to an emerging category of communications devices that will include small, lightweight multifunction portable phones, portable facsimile and other imaging devices, multichannel cordless phones, and advanced paging devices with two-way data capabilities.

PDA An acronym for personal digital assistant, a class of handheld devices equipped with onboard processing power. Most PDAs are designed to run personal organization software (contact management, calendar, etc.) and some light business productivity applications. Most PDAs offer wireless connectivity.

Perl Practical Extraction and Report Language, a popular scripting language used for all kinds of processing tasks on the Web, including forms processing.

PGP Pretty Good Privacy (PGP) is one of the most effective data encryption schemes available today. It generates two keys that belong uniquely to you. One PGP key is secret and stays in your computer. The other key is public and is given out to people you want to communicate with. The key you distribute enables others to decrypt your messages back into their original form.

phoneware A casual term for software that enables users to make telephone calls over the Internet.

PIN Pager Identification Number (PIN) is a unique number assigned to a pager, allowing it to receive messages addressed to the subscriber of the paging service.

ping A ping is a network packet which is sent from one computer to another and then echoed back for the purpose of measuring packet delay (in milliseconds) and packet loss between two hosts on the network. This is done by using a ping utility that sends a signal at regular intervals to the remote machine which essentially asks, "Are you there?" If the remote machine is turned on and is able to send and receive messages, it will respond to the ping's inquiry. If there is no response, this means the remote machine is not functioning or a firewall has blocked the ping signal. In either case, you will not be able to communicate with the desired IP address. (See also **trace route.**)

plug-in A plug-in is software that gives a Web browser extra features. Upon installation, the plug-in sets itself up as part of the Web browser. For example, there are VRML plug-ins that allow Netscape Navigator to display three-dimensional images written in the Virtual Reality Markup Language.

pong A reply to a ping message.

POP Server A server that uses the Post Office Protocol, which holds incoming e-mail until the recipient is ready to read or download it.

PPP Point-to-Point Protocol is used to establish connections to remote networks via a PC's COM port and modem. When the user dials into a remote network, the PC becomes a node on that network. PPP supports multiple types of data transmission protocols, such as TCP/IP and IPX/SPX. PPP is expected to supplant SLIP as the preferred method of establishing dial-up connections to the Internet. PPP provides automatic negotiation of the network configuration without user intervention and supports encrypted authentication.

protocol   A set of rules and conventions for exchanging data over a network. The term also refers to the software and/or hardware implementation of these rules.

proxy server   A type of server that offers performance improvements by using an intelligent cache for storing retrieved documents. The proxy's disk-based caching feature minimizes use of the Internet by eliminating recurrent retrievals of commonly accessed documents, caching them locally. This feature improves interactive response time for locally attached clients. The resulting performance improvements provide a cost-effective alternative to purchasing additional network bandwidth. Proxy servers also eliminate the need to assign an IP address to every desktop, which reduces administrative time and expense. Allowing users to share a single IP address also helps tighten the security of the corporate network.

Pulse Code Modulation   PCM is an encoding technique used by carriers to change voice signals into digital form for transmission over the telephone network. Under PCM, voice signals are sampled at the minimum rate of two times the highest voice frequency level of 4000 Hz. This translates into a rate of 8000 samples per second. The amplitudes of the samples are encoded into binary form using enough bits per sample to keep the quantizing noise low, while maintaining a high signal-to-noise ratio.

RAM   Random access memory, memory that temporarily holds frequently accessed data needed to run applications. Most computers sold today come with a minimum of 8 MB (megabytes) of RAM, which is enough to support most Windows-based applications.

reflector   Server-based software that allows users to meet at a single point to conduct and manage meetings over the Internet, private TCP/IP networks, and the MBONE. The reflector provides such advanced features as network bandwidth control, video pruning, audio prioritization, and conference management. (This software is also known as a *director.*)

relative URL   A reference to a document or service on the same server. (See also **absolute URL.**)

remote access There are two remote-access solutions: remote control and remote node. With remote-control software the user dials up a specific modem-equipped or LAN-attached PC to access its files, download e-mail, or troubleshoot a problem. As long as the software is installed on both ends, the remote computer assumes the capabilities of the host. Everything on the host's screen is mirrored on the remote computer's screen. With remote node, the user's modem-equipped PC dials into the LAN and behaves as if it were a local LAN node. Instead of keystrokes and screen updates, the traffic on the remote node's dial-up line is essentially normal network traffic. The remote PC does not control another PC, as in remote control; rather, it runs regular applications as if it were directly attached to the LAN.

RSVP An Internet control protocol, the resource ReSerVation Protocol runs on top of the Internet Protocol (IP) to provide receiver-initiated setup of resource reservations on behalf of a multimedia application data stream. When an application requests a specific quality of service for its data stream, RSVP is used to deliver the request to each router along the path(s) of the data stream and to maintain router and host states to support the requested level of service. In this way, RSVP essentially allows a router-based network to mimic the circuit-switched network on a best-efforts basis. (See also **RTP.**)

RTP The Real-time Transport Protocol is typically implemented as part of the application. RTP provides functionality suited for carrying real-time content, specifically, a time stamp and control mechanisms for synchronizing different streams with timing properties. RTP inserts timing and sequencing information into each packet. Although RTP does not enhance the reliability of a transmission, applications make use of the timing and sequencing information to enable audio and video streams to be played smoothly, despite occasional packet loss.

sampling A method for converting analog speech into digital form. A high sampling rate results in high-quality reproduction but requires more system resources to implement.

script A series of commands that perform a specified function, such as establish a connection or process a form.

server A computer that handles requests for data, e-mail, file transfer, and other network services. A server at a remote site is often called a *host*.

server-side includes The ability of the Web server to automatically modify a file as it is being served to include information from other files or environment variables, such as date and time.

SGML Since 1986, the Standard Generalized Markup Language is an ISO standard for document markup for both print and online publication. HTML 2.0 is based on some of the concepts of SGML. HTML 3.0 will be fully compliant with SGML.

SLIP Serial Line Internet Protocol is used to establish connections to remote networks via a PC's COM port and modem. When the user dials into a remote network, the PC becomes a node on that network. Whereas SLIP supports only TCP/IP, PPP supports multiple data transmission protocols. Unlike PPP, SLIP does not provide automatic negotiation of network configuration without user intervention and does not support encrypted authentication.

SMTP server A server that uses the Simple Mail Transfer Protocol to send and route e-mail over the Internet.

SNPP A gateway protocol, Simple Network Paging Protocol (SNPP) speeds message delivery from the Internet to commercial paging networks and, unlike e-mail, sends an immediate confirmation that the data has been received.

spamming A casual term that refers to the unsolicited broadcast of messages to newsgroups or e-mail boxes, mostly by businesses or individuals trying to sell products and services.

SSL The Secure Sockets Layer (SSL) protocol safeguards communication between clients and servers on the Internet. SSL provides encryption, server authentication, and message integrity, which address the security concerns of companies and individuals.

static IP address With static IP addressing, a permanent IP address is assigned to each machine that is connected to the Internet.

syntax The syntax of an HTML document defines what is permissible in terms of the tags, names, attributes, and values that describe document elements.

T.120 T.120 is a suite of dial-up and network-based data conferencing protocols recommended by the International Telecommunications Union (ITU). T.120 specifies how applications may interoperate with (or through) a variety of network bridging products and services that also support T.120.

Tags In HTML documents, tags define the start and end of headings, paragraphs, lists, character highlighting, links, and other document elements. Most HTML document elements are identified with a start tag, which gives the element name and attributes, followed by the content, and then the end tag. Start tags are encapsulated by < and >, while end tags are encapsulated by </ and >.

TCP/IP Transmission Control Protocol/Internet Protocol, the de facto protocol suite for data transmission over the Internet.

Telnet A type of connection that is used to establish connectivity between a host and a client computer over the Internet, usually requiring a user name and password. Telnet access enables the user to type commands which are executed at the host. The results are passed back to the client computer. Among other things, these hosts often include archives, chat systems, and games.

TIFF Tagged Image File Format, a common format for storing color and gray-scale images, which is not supported by Web browsers.

trace route A program activated within a ping utility that allows you to chart a path between two hosts on the Internet. The trace route program reports each Internet router hop and the time it took for your test to get there. If you see a high number of milliseconds (ms) for some of the hops, this indicates where the problems are in your connection. (See also **ping.**)

TrueSpeech A product of the DSP Group, TrueSpeech provides upward of 18:1 compression of raw audio with an imperceptible loss of audio quality on decompression.

UDP User Datagram Protocol, a streamlined datagram service in which there is no facility for detecting the delivery or duplication of datagrams (i.e., packets). The datagrams are passed to IP with very low overhead (just a source and destination port, length, and checksum) in the header. Since UDP uses IP, it is a connectionless service which is inherently unreliable in that it does not use acknowledgments, control the order of arrival, or provide flow control.

unicast A method of information delivery in which each packet is duplicated x times, where x equals the number of recipients. This taxes network bandwidth and your computer's or server's CPU. (See also **MBONE** and **multicast.**)

UNIX A multiuser, multitasking 32-bit operating system that is used by most Web servers. UNIX was originally developed by AT&T Bell Laboratories.

upload The process of transferring files from the local computer to a remote system over the network.

URL Uniform Resource Locator, a string of characters that uniquely identifies each page of information on the World Wide Web. An example is http://iquest.com/~nmuller/index.html.

Usenet An electronic forum on the Internet of several thousands of different topical discussion groups or newsgroups.

UU A document that is UNIX-to-UNIX (UU)-encoded is packaged for transmission in a way that can be understood by the network. Upon delivery, the document is UU-decoded to its original form so it can be understood by the application.

UUCP UNIX to UNIX Copy Protocol, a method of transferring files between computers, including electronic mail. An alternative is the Simple Mail Transfer Protocol. (See also **SMTP.**)

variable In HTML, the portion of a tag that a user can change to influence how text or images in a document are rendered by the browser.

Veronica A client-server system that provides a way to search for a key word in all Gopher menus at all Gopher sites known to the Veronica server database.

video mail The capability to record a video message for delivery via e-mail as a MIME attachment.

Video-on-Demand A method of video distribution that entails the storage of video files at a server in compressed form that can be accessed on demand by authorized users with the proper viewer. VOD files can consist of announcements, product demonstrations, training modules—virtually anything a person or a company has recorded and stored for future reference.

viewer A specialized program that is launchable from a browser, which allows the user to access multimedia files, such as audio and video clips.

VRML Virtual Reality Markup Language is a set of tags that allows Web page developers to design three-dimensional objects and environments. Participants can collect objects and interact with each other as they move through these environments. The environments can be cybermalls, virtual offices, or fantasy worlds.

W3C World Wide Web Consortium, created by the Massachusetts Institute of Technology (MIT) and CERN to direct the development of the Web. When CERN pulled out of further Web development in 1995, it appointed INRIA, France's national computing research institute, to take its place.

WAIS Short for *Wide Area Information Service*, a network service that is used to search for information by key words or phrases in specially indexed files.

WAN An acronym for *wide area network*, which connects computers over long distances via carrier-provided lines and services.

Web server A computer connected to the World Wide Web network for disseminating information to individuals browsing the Internet.

Web spew The universe of useless news, opinion, and other wasteful verbiage routinely published on the World Wide Web as if it were really important.

Webmaster A person who manages a Web site and who also may have responsibility for retrieving and responding to e-mail messages concerning the content of the site's Web pages. Whether or not the Webmaster actually manages the Web server, the term implies a high degree of expertise in HTML, forms, and associated programs. The term is believed to have been coined by Hassan Schroeder who works at Java-Soft, a Sun Microsystems' company.

whiteboard A facility offered by chat programs and collaborative computing programs that allows conference participants to draw in real time and view each other's work. A whiteboard often includes many of the basic rendering tools found in graphic design and technical drawing products.

Windows NT A 32-bit operating system developed by Microsoft Corporation. It comes with preemptive multitasking, strong networking support, no memory limits, and no dependency on DOS. Windows NT claims equivalent power to UNIX.

Winsock Short for *Windows Sockets*, a commonly used protocol stack, Winsock implements TCP/IP network functions within Windows applications.

wireless IP Another term for Cellular Digital Packet Data (CDPD), a wireless data network built on top of the nation's cellular telephone infrastructure. CDPD transmits information using the Internet Protocol (IP).

WWW World Wide Web, also known as the Web, is a service on the Internet that weaves information and resources together through the use of hypertext links.

XX XX is a document encoding/decoding scheme that works similar to UU, but uses a different character set so that character set translations will work better across multiple types of systems—between IBM's mainframe-oriented EBCDIC and ASCII, for example.

INDEX

Page numbers appearing in italics refer to figures.

A

E

Index

X

Y

Z

ABOUT THE AUTHOR

Nathan Muller is an independent consultant in Huntsville, Alabama, specializing in advanced technology marketing and education.

In his 25 years of industry experience, he has held numerous technical and marketing positions with such companies as Control Data Corporation, Planning Research Corporation, Cable & Wireless Communications, ITT Telecom, and General DataComm. He has and M.A. in Social and Organizational Behavior from George Washington University.

Muller has written extensively on many aspects of computers and communications, having published over 1000 articles and 12 books on such diverse topics as frame relay, the synchronous optical network, LAN interconnection, intelligent hubs, network management, document imaging, wireless data networking, and the Internet.

Muller is a regular contributor to the technical reports published by Datapro and Northern Business Information, two of The McGraw-Hill Companies, as well as *BYTE Magazine*, which is published by McGraw-Hill. He also is on the editorial board of the *International Journal of Network Management* published by John Wiley & Sons.